The Great Transformation

How Contemporary Science Harmonizes
with the Spiritual Life

George Stanciu

Principle Source Publisher

The author wishes to thank numerous colleagues and students for engaging in intense, fruitful dialogues that repeatedly confirmed the strange, amazing depth to things that goes well beyond the "truths" of common sense or cultural formation.

Table of Contents

Introduction: The Kali Yuga

Like you, I was born in the Kali Yuga, the Dark Age of Hindu mythology, when all the great religious faiths of the world are on the wane.

The secular faith in the Nation-State, in grand schemes to institute Paradise on Earth, and in placing transcendent hope in human institutions has been destroyed by history. No theoretical arguments are necessary to show that the goal of a Heaven on Earth is perverse and that the pursuit of such a goal leads to untold death and destruction, to a Hell on Earth for tens of millions.[1] The sight of the rubble of Hiroshima, the smell of burning bodies in Auschwitz, and the sound of frozen corpses thrown on sledges in the Gulag destroyed secular faith. A technological utopia, a Master Race, and a classless society are nightmares from the past, only believable to a handful of science-fiction writers, to a few crazy ideologues blind to history, and perhaps to one or two drunks in bars near Harvard and M.I.T.

Secular faith is dead, and religious faith is stumbling toward the graveyard. Pope Benedict XVI, in the first year of his pontificate, lamented the weakening of churches in Europe, Australia, and the United States. "There's no longer evidence for a need of God, even less of Christ," he told a meeting of clergy in the Italian Alps. "The so-called traditional churches look like they are dying."[2] Capitalism focused attention on material prosperity, on the good life in this world, away from eternal salvation, so that few Christians today see themselves as pilgrims journeying through this Earthly life, shunning the attachment to worldly things and avoiding the snares set by Satan.

Not surprisingly, the buried stream of nihilism running for a century or more beneath Modernity surfaced in the twentieth century and is now a swift river, carrying many away into moral and intellectual chaos. Music, poetry, philosophy, religion, and morality are believed to be mere expressions of personal opinions, individual perceptions, or particular cultural viewpoints.

In America, the authority of every institution over the last fifty years weakened to nonexistence. The Church does not command respect, nor does the United States Congress, so neither religious truths nor a common good can serve as the basis for an argument. In addition, custom eroded to the point that anything is permitted provided no innocent bystander is injured. Under the increased sway of democratic equality, schools and colleges no longer inculcate values or even social etiquette.

Traditionally, for some people, philosophy furnished universal truths about the human being, but those happy days disappeared years ago. Martin Heidegger, probably the most influential philosopher of the twentieth century, declared in a posthumously published interview with *Der Spiegel*, "Philosophy is at an end."[3] Friedrich Nietzsche, of course, was there first. In an 1873, unpublished essay, "On Truth and Lie in an Extra-Moral Sense," he asked, "What is truth?" and answered, "A mobile army of metaphors, metonyms, and anthropomorphisms."[4]

In his book *Consequences of Pragmatism*, philosopher Richard Rorty envisages a post-philosophical culture, where the search for ruling principles, immovable centers, and fixed structures is abandoned. According to Rorty, historians, literary theorists, and philosophers in the future will accept that no escape from culture exists and that all intellectuals can do is ride the "literary-historical-anthropological-political merry-go-round," chasing one intellectual fad after another. In a world without maps, in a culture without fixed reckoning points, "there is nothing deep down inside us except what we have put there ourselves."[5]

The only intellectual discipline that a thinker can draw upon is science, for in our chaotic times, science is still widely believed to be the only path to truth, a belief that killed theology, philosophy, and poetry. We must not be so naive to believe that science exists in some pure land. When a scientist steps into the laboratory or sits down at a desk to theorize, he or she is not suddenly transported to an ethereal region, high above the rest of humankind. The scientist, just like the poet, the artist, and the philosopher, is reared in a community with a specific understanding of the world and thus brings to science hidden cultural assumptions about mind, matter, and what constitutes knowledge. Culture often poses the questions we ask, the facts we seek, and the kind of answers we accept. Consequently, the results of modern science are habitually interpreted through culture.

Nevertheless, science often transcends culture to universal truths. In contrast to Western civilization's descent into political and intellectual chaos in the twentieth century, science underwent an unexpected and welcoming revolution that stripped away the cultural blinders of materialism and threw open the door to the nonmaterial with the breath-taking vision that the spiritual life could be supported by empirical science. *The Great Transformation* uncovers our spiritual nature by first clearing away the cultural obstacles on our only path to truth and then laying out how in contemporary science the cosmos is composed of two elements, mind and matter, neither reducible to the other.

1 A New Science, A New Cosmos

In our consumer society, advertisers use the powerful word "new" to sell products because many buyers want the current cell phone, the latest fashions, and the most recent expresso machine, even if none of these differ substantially from the older ones.

We avoid the word "new" in our discussion of the eras of science; instead, we adopt the terminology used to describe natural phenomena. For us, the change from a horse-drawn buggy (mobile) to a Ford Model T (automobile) is a phase change, from one different form of transportation to another, like ice melting to liquid water, each with its own distinct properties. In 1951, the Ford Victoria was available for the first time with Ford-O-Matic, an automatic transmission. The change in transmissions we call developmental since the form of transportation was not changed, merely improved.

Unlike the melting of ice to liquid water, historical phase changes cannot be reversed. We cannot go home again to the cozy, ancient cosmos, where the night sky displayed the transcendent and Mother Earth manifested harmony and fecundity.

Figure 1.1. Time exposure with the camera pointed to the North Star.

The Aristotelian Cosmos

Aristotelian science rests on the careful observation of common experience. You can go outside at night, like Aristotle (384–322 BC) and Ptolemy (c.AD 100 – c.170) did, sit on the ground, and see the stars trace out circles around an imaginary axis that goes through the North Star and the Earth. Unless you are a human being unique in history, I can guarantee you will not feel that you are in motion but that the stars are turning around the Earth. (See Figure 1.1.) Repeated viewing of the stars over many years shows that the constellations and other stars do not change in brightness or in location with respect to each other. The constellation Pisces looked the same to Aristotle and Ptolemy as it does to us.

The stars, clearly, move differently than terrestrial objects: A rock held up in the air and released falls in a straight line toward the center of the Earth. From such common experience, Aristotle drew an obvious conclusion: The cosmos is composed of two kinds of matter, celestial and terrestrial. The celestial matter of the stars is eternal, moves in perfect circles, and is either divine or moved by something divine. In contrast, the Earth is composed of ever-changing matter whose natural movement is in a straight line toward the center of the Earth.

In the Aristotelian-Ptolemaic Cosmos, the Earth is the basement, the sinkhole where all the gross, dull matter is concentrated. Renaissance philosopher Giovanni Pico della Mirandola described the dwelling place of Earth as "the excrementary and filthy parts of the lower world."[6] Inhabitants of the cosmic sinkhole suffered floods, earthquakes, and plagues, while the angels and saints in heaven blissfully contemplated God. The great French essayist Michel de Montaigne said people feel and see themselves "lodged here amid the mire and dung of the world, nailed and riveted to the worst, the deadest, and the most stagnant part of the universe, on the lowest story of the house, and the farthest from the vault of heaven."[7] The Earth's place in a geocentric cosmos is not preeminent, nor does it lead to human pride or naïve self-love, unless the bubonic plague is preferable to eternal bliss.

Following Aristotle, Ptolemy begins his great work on astronomy, the *Almagest* (c. 150 AD), by first laying out in broad outline the three categories of knowledge — physics, mathematics, and theology. "In the highest reaches of the universe, completely separated from perceptible reality," is an invisible and motionless deity (the Prime Mover), who is the "first cause of the first motion of the universe."[8] The first cause is the most intelligible entity that exists. Never changing, the Prime Mover is unknowable to us because of its remoteness. Properly speaking, theology should be called guesswork, not a science, since the invisible and motionless deity is remote and unknowable.

Physics, the opposite of theology, investigates what is perceptible and near at hand. Terrestrial matter, unfortunately, is unstable and obscure, for such qualities as white, hot, sweet, and soft never remain. What is closest to us is least intelligible and thus, on the whole, unknowable. Like theology, physics is guesswork, but for a different reason; terrestrial matter is essentially unknowable. Consequently, philosophers can have no hope of agreeing about either the Prime Mover or terrestrial matter. Only mathematicians using the indisputable methods of arithmetic and geometry have arrived at sure and unshakeable knowledge because mathematics occupies a middle ground between theology and physics. Numbers and geometrical figures are close at hand yet unchanging.

Occupying the basement of the universe, the sinkhole of the cosmos, the ancient astronomer is limited in what he can understand. The night sky is an impenetrable mystery, and the astronomer can offer only likely stories about the motion of the planets, not explanations. At no place in the *Almagest* does Ptolemy give a drawing of the cosmos; each planet is a divine being that does its own thing, and Ptolemy has no reason to think that Mercury, Venus, Mars, Jupiter, and Saturn fit together into a grand scheme.

Biology

Aristotle grasped that the central problem of biology is to understand the bewildering variety of organic forms. A glance at nature reveals dandelions, ferns, orchids, aspens, junipers, worms, beetles, mayflies, fleas, minnows, swordfish, whales, armadillos, hawks, hummingbirds, bats, mice, squirrels, rabbits, wolves, bears, a seemingly unending number of living things with no apparent connection to each other. Aristotle created the first intellectual tool for biology, the method of collection and division, used in one form or another by subsequent biologists to group organisms into related categories. In Aristotle's classification scheme, the genus animal includes all organisms that perceive and move themselves. The genus animal is then divided into mutually exclusive groups. The definition "rational animal" sets Homo sapiens apart from all other animals. In modern classification schemes, nature is divided into living and nonliving; then, the broad category living is further divided into domains, kingdoms, phyla, classes, orders, families, genera, and species, an example of developmental change in science that extends over 2,400 years.

The method of collection and division provided a weak unification of living things. Aristotle and his students collected through direct observation a compendium of botanical and zoological facts published in the treatise *On the Parts of Animals*, a fascinating gathering of one damn thing after another: Of all animals, man has the largest brain in proportion to his size; dolphins have bones not fish-spines; the Libyan ostrich is an exception among birds for it has eyelashes; and on and on.[9]

In search of a greater unity of organisms, Aristotle made the commonsense assumption that human beings are as much a part of nature as dandelions and hawks; consequently, how humans and nature produce things are in principle the same: "For just as human creations are the products of art, so living objects are manifestly the products of an analogous cause or principle, not external but internal."[10] Or said in a formal, mathematical way, (products of art):art :: (products of nature):nature.

To understand nature, Aristotle directs us to first go into the workshop of any craftsman, say a potter, and observe what he does. The potter throws clay on a table, shapes the clay with his hands into some form, say a jug to hold water. Careful observation reveals that the potter making a jug exhibits four causes: the clay (the material cause), his hands (the efficient cause), the shape of the jug (the formal cause), and that for-the-sake-of-which the action was done, to make a jug to hold water, the purpose (the final cause). Nature, too, exhibits these four causes. An acorn desires to grow into a fully mature oak tree (final cause); the acorn absorbs water and nutrients from the soil and light from the sun (the material cause); the material is transformed into a growing oak (the efficient cause); the formal cause is the shape of the mature oak tree, say the characteristic oak leaves, bark, and limbs.

Aristotle introduces the final element that unifies astronomy and biology — desire. The stars, made of divine, incorruptible matter, desire to be like the Prime Mover and thus move themselves in perfect circles. Every plant and animal desires full actuality like the Prime Mover has and that causes seeds and fertilized ova to grow into adult organisms. Each plant and animal desires immortality, but the instability of matter prevents that, so a plant or an animal settles for what is second best, reproducing itself to continue its species, and that explains the cycles of generation and corruption found throughout nature. The source of every natural motion is an agent desiring some end. Even terrestrial matter has an agency; a rock falls to the ground because it desires to be at the center of the Earth. The universal principle of the Aristotelian Cosmos is not material causation but the Prime Mover that "produces motion as being loved."[11]

In the Aristotelian Cosmos, *Homo sapiens* is a part of nature and shares a common nature with plants and animals. Like plants, human beings respire, require nourishment, and reproduce. Like animals, human beings perceive, move themselves, and, like the higher animals, have emotions. What makes *Homo sapiens* unique in the natural world is an intellect and a will; our thinking emulates the Prime Mover's thinking upon thinking.

Aristotelian science accounted for all human experience[12] until the phrase change introduced by Copernicus.

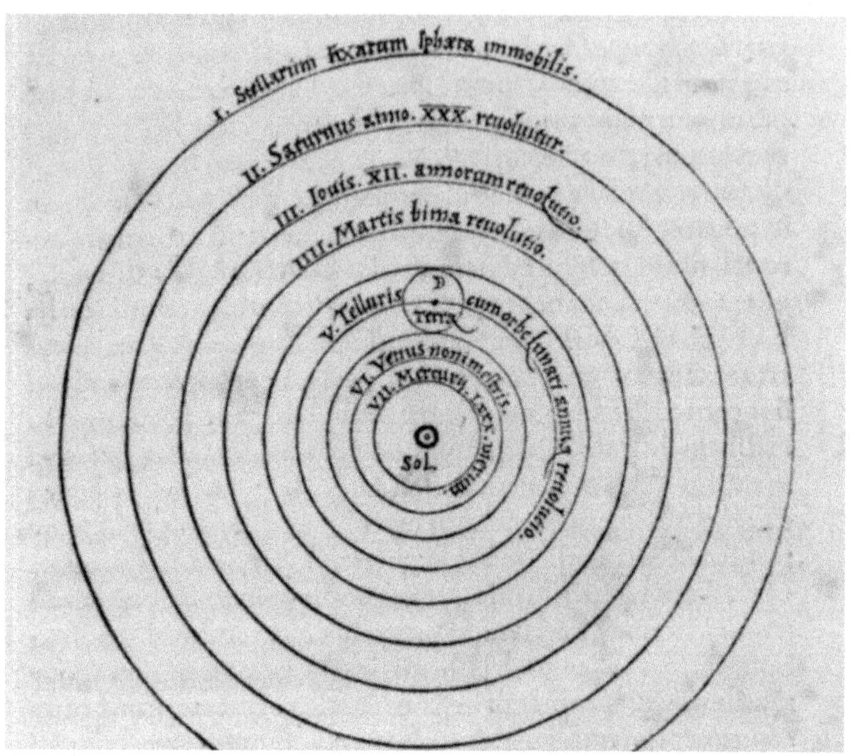

Figure 1.2. To persuade his reader of the truth of the heliocentric cosmos, Copernicus offered as overwhelming proof the harmony and proportion of such a universe when seen from the outside, from God's perspective. (Copernicus, 1543)

The Copernican Revolution

Nicolaus Copernicus (1473-1543) begins astronomy where Aristotle and Ptolemy could not, with the book of Genesis. From Revelation, Copernicus concludes that the cosmos was fashioned by "the Best and Most Orderly Workman of all."[13] Consequently, the cosmos must be intelligible and beautiful. Copernicus' main critique of Ptolemy and his followers is that they did not discover the chief point of astronomy, the beautiful form of the cosmos. The geocentric cosmos lacks unity, and the parts are not harmonious. Copernicus likens a Ptolemaic astronomer to an artist taking hands, feet, head, and limbs from his different drawings of humans, each part beautifully drawn, and assembling them together; the result is a monster, since no two parts match. From Genesis, Copernicus concludes that the Aristotelian-Ptolemaic Cosmos must be wrong.

The second principle that Copernicus draws out of Genesis is that since man and woman are made in the image of God, natural philosophers can discover "the truth in all things, in so far as God has granted that to human reason."[14]

In Book One of *On the Revolutions of the Heavenly Spheres*, Copernicus hurries to persuade the reader of the truth of the heliocentric cosmos; he offers as overwhelming proof, the harmony and proportion of such a universe when seen from the outside, from a divine perspective, see Figure 1.2. Looked at from the outside, from God's vantage point, the cosmos is seen to have a wonderful harmony and simplicity. The astronomer sees the truth of the heliocentric cosmos by simply looking at it. Copernicus exclaims, "How exceedingly fine is the godlike work of the Best and Greatest Artist!"

Unlike Ptolemy, Copernicus draws a map of the cosmos. The astronomer sees that in a heliocentric cosmos, the Earth is a planet, a wandering star: "The center of the Earth too traverses that great orbital circle among the other wandering stars in an annual revolution around the Sun."[15] No longer are the stars remote; we are riding on one in our annual traverse around the Sun! The Copernican Revolution unnails and unrivets us from sinkhole Earth.

From God's vantage point, we see that the Aristotelian division of matter into celestial and terrestrial is wrong. The Earth is a wandering star, so either the Earth is divine or the stars are mundane like the Earth. The Earth, clearly, is mundane; thus, the stars must be too. In the Copernican Cosmos, the planets and the Earth move in perfect circles. Somehow the mundane matter of the Earth must have regular mathematical properties and thus must be intelligible. Therefore, the intelligible stars can be known to us. Star-stuff and Earth-stuff are the same; we can study star-stuff by going out in the backyard and digging up earth. No longer must the scientists be satisfied with accounts, with likely stories. The Copernican Revolution is a phase change in science: For the first time in human history, scientists can explain how things *really* are.

The truth of the heliocentric cosmos gave rise to a pressing problem. The mathematical properties of a shovelful of earth are not obvious, so a new science was needed, a new method to replace Aristotelian science.

Galileo (1564-1642) grasped three key elements of the new science. Although his telescope revealed that moons orbit Jupiter and thus offered evidence for the truth of the Copernican system of the Sun and planets, his inclined-plane experiment to determine how terrestrial bodies accelerate touched nature deeper. For the first time in history, Galileo showed that terrestrial matter obeys mathematical laws, a more powerful confirmation of the heliocentric universe than the observation of the moons of Jupiter. This confirmation was done through experiment, not observation.

In addition, Galileo enunciated with great clarity a basic assumption of the Copernican Revolution and thus modern science: "The grand book [of] the universe, which stands continually open to our gaze. . . . is written in the language of mathematics, and its characters are triangles, circles, and other geometric figures without which it is humanly impossible to understand a single word of it; without these, one wanders about in a dark labyrinth."[16] Today, the grand book of nature is written in the characters of tensors, group theory, and Hilbert spaces.

Finally, Galileo declared that the human being was not an integral or representative part of the universe: "I think that tastes, odors, colors, and so on are no more than mere names. . . . They reside only in the consciousness. Hence if the living creature were removed, all these qualities would be wiped away and annihilated."[17] The goal of new science is to understand nature in the absence of the scientist.

Although Galileo performed the first experiment, he did not fully understand, as Francis Bacon (1561-1626) did, the difference between observation and experiment. Like Copernicus, Bacon began with Genesis. He argued that human knowledge had fallen into decay: In the Garden of Eden, Adam named the animals according to their nature, and this knowledge gave Adam command over nature. "The state of knowledge is not prosperous nor greatly advancing," Bacon lamented and declared, "a way must be opened for human understanding entirely different from any hitherto known."[18] The main impediment to the advancement of knowledge was Aristotelian philosophy filled with "specious and flattering" propositions that gave rise to "contentious and barking disputation."[19] To partially restore mankind to the Garden of Eden, the only time in human history that man had real authority over nature, a new science was needed "in order that the mind may exercise over the nature of things the *authority* which properly belongs to it."[20]

Bacon was the first person to enunciate the fundamental principle of the new science: "The testimony and information of the sense has reference always to man, not to the universe; and it is a great error to assert that the sense is the measure of things [as Aristotle did]."[21] But a total rejection of the senses is madness, so to arrive at trustworthy information about nature the senses must be assigned a limited role. In one sentence, Bacon presented the heart of the experimental method, something entirely new to mankind: "The office of the sense shall be only to judge of the experiment, and the experiment itself shall judge of the thing."[22] Said another way, the scientist touches the experiment, and the experiment touches nature. The scientist has no direct contact with nature.

Bacon introduced one other principle entirely new to mankind: The true test of human knowledge is whether nature can be commanded, for "those twin objects, human knowledge and human power, do really meet in one; and it is from ignorance of causes that operation fails."[23] While the goal of the new science is to "command nature in action," the goal of Aristotelian philosophy was "to overcome an opponent in argument."[24] Bacon tossed Aristotle on the trash heap of history, for implicit in Aristotelian physics is that terrestrial matter is obscure and barely knowable and thus cannot be commanded.

The new science would make mankind the master and possessor of nature, much as Adam was in the Garden of Eden, a time when "man was immortal and able to meet God face to face."[25] Assuming the Genesis story is consonant with other myths of Paradise, we can speculate that Adam was "happy, and did not have to work for his food: Either a Tree provided him with subsistence, or the agricultural implements worked for him of themselves, like automata."[26]

In effect, Bacon rewrote the Bible. The Old Testament told how Adam and Eve were expelled from the Garden of Eden; the New Testament gave hope that what had been lost through the disobedience to God would be restored at some point through faith in Jesus Christ. In Bacon's new version of Genesis, the restoration of the lost Paradise was expected to come about not through faith but from the "great mass of inventions"[27] that would flow forth from the new experimental science that would give mankind the command over nature that Adam had in the first Garden of Eden. As a result, the descendants of the first man and woman could on their own return to Paradise, and in this way "subdue and overcome the necessities and miseries of humanity"[28] that resulted from the expulsion of Adam and Eve to East of Eden, where women painfully suffered childbirth and the cursed ground brought forth thistles and thorns.

Bacon, with great clarity, introduced a phase change in science, technology, and human aspirations: Science founded on experiment produces human knowledge and human power that together improve the lot of mankind. With this phase change, there was no going back to Aristotelian science.

René Descartes clarified the remaining fuzziness in the foundations of the new science. Like Galileo, Descartes held that to know nature was to know it mathematically. He reasoned mechanics was mathematical; therefore if the universe was intelligible, "the laws of mechanics are identical with those of nature,"[29] and the mechanical universe was born. His idea of order was characterized by regularity, necessity, and mathematical certainty as exhibited by machines, and his symbol of such order was the clock. Descartes envisaged the intelligible physical world as a huge clockwork mechanism buried beneath the world of appearances; the rational structure of the universe was a mechanism that operates according to unchanging mathematical principles. In contrast to organic order which seems purposeful and human activity, which is almost always purposeful, mechanical order has the peculiarity that there is no purpose, no guiding direction to the mechanical activity of nature. An idiot machine runs without purpose, operating by necessity: Weights fall, a drive mechanism turns, causing the engaged gears to go around — a chain of mechanical causation and at no time does purpose, desire, or will enter the workings of the machine.

The Thirty Years War and the harsh winter of 1619 forced Descartes to take up temporary quarters in Germany: "I remained the whole day shut up alone in a stove-heated room, where I had complete leisure to occupy myself with my own thoughts."[30] Descartes meditated upon the disunity and uncertainty of the knowledge of his day. He marveled at mathematics and wondered why the certainty and evidence of its arguments had not been used as the foundation for all knowledge. Unlike philosophers, mathematicians possessed genuine knowledge, for the demonstrations of geometry were so evident and clear that no one disputed them. Then, in a blinding flash, Descartes conceived of a method to put all knowledge on a footing as secure as that of mathematics. He explained that the essence of his new method is to commence "with objects that were the most simple and easy to understand, in order to rise little by little, or by degrees, to a knowledge of the most complex."[31]

Now known as reductionism, the Cartesian method has two steps: first thoroughly understand the simplest parts, and then construct the whole as a sum of the parts. Descartes believed the reason mathematicians give such certain demonstrations is that they follow the Cartesian two-step procedure.

The Newtonian Cosmos

It fell to Newton, not Descartes, to discover the clockwork mechanism. In 1687, Isaac Newton firmly established the mechanical universe when he published *Philosophiæ Naturalis Principia Mathematica*, often called the *Principia*, a work that changed science and the world forever. Newton purported to discover a huge clockwork mechanism governed by unchanging mathematical principles beneath the ever-changing world of physical appearances. In Newtonian mechanics, the world is made up of forces and masses (matter); the mechanical forces of pushes and pulls cause changes in the motion of masses.

In the preface to the *Principia*, Newton sets out the goal of modern science: "I derive from the celestial phenomena the forces of gravity with which bodies tend to the sun and the several planets. Then from these forces, by other propositions which are also mathematical, I deduce the motions of the planets, the comets, the moon, and the sea. I wish we could derive the rest of the phenomena of Nature by the same kind of reasoning from mechanical principles, for I am induced by many reasons to suspect that they may all depend upon certain forces by which the particles of bodies, by some causes hitherto unknown, are either impelled toward one another and cohere in regular figures, or are repelled and recede from one another."[32] The Newtonian method, then, aims to explain the behavior of a complex system in terms of its simpler parts and their interactions, and thus is Cartesian reductionism stated in terms of mechanics. In the Newtonian map of nature, only matter moved about by mindless forces exists.

For Enlightenment thinkers, Newtonian physics was the newly discovered key to understand nature, mankind, and God. The universe was a giant clockwork mechanism, man a machine, and God the Designer of all. Thomas Jefferson, while ambassador to France, commissioned a portrait of the new holy trinity in which Descartes and Bacon each occupied a bottom vertex of an equilateral triangle, while Newton surveyed the cosmos from its apex. In the Enlightenment, Newtonian mechanics became the new truth that encompassed all existence.

With Newton, the phase change in science was fully solidified. The mathematical nature of the universe is astounding. Except for Pythagoras' vague, mystical theories and Plato's speculation that the universe is composed of the five Platonic solids and musical harmonies, the ancients, including Aristotle and Ptolemy, never envisioned that terrestrial phenomena obey mathematical laws. Earthly motion is bafflingly complex, and unlike the Sun and Moon, the prediction of the future of anything on Earth seems impossible. It is far from evident that a projectile on Earth traverses a parabolic path, that the colors of the rainbow exhibit a simple mathematical pattern, and that underlying the bewildering, everyday world of chemical changes, such as the rusting of iron or the burning of sugar, are simple, mathematical rules. Every herder and farmer, since the domestication of animals and the inception of agriculture, knew an offspring resembles its parents, but Gregor Mendel was the first person to discover that the laws of inheritance are written in terms of simple, whole numbers.

The Newtonian Cosmos would last for nearly four hundred years. Knowing nature through mathematics and experiment led to the intellectual jewels of modern Western culture: Newtonian physics, Maxwell's electrodynamics, special and general relativity, quantum physics, and the discovery of the physical basis of life, including the genetic code.

As we will see, in the twentieth century, quantum physics, chaos theory, and neuroscience bought about a new phase change in science. The mechanical universe was relegated to the Dark Ages of Science; determinism, reductionism, and materialism, nightmares of the past, vanished. Dead matter, the universal cause of all existence, was replaced by the primacy of mind.

2 Dumb Idea #1: Determinism

Despite the advent of relativity, quantum physics, and chaos theory, most scientists, including most physicists, intellectually inhabit the Newtonian Cosmos. In stark contrast to the Aristotelian Cosmos, where plants and animals possess an inner agency that causes them to emulate the Prime Mover, the Newtonian Cosmos is mechanical, where lifeless matter, as well as animate beings, is moved solely by pushes and pulls. The Newtonian clockwork universe obeys precise, mathematical laws that rigidly determine its evolution; consequently, free will in this cosmos is impossible.

In his Philosophical Essay on Probabilities, published in 1814, French mathematician Pierre Simon Laplace made explicit the determinism that would dominate the thinking of scientists for the next two hundred years: "We ought then to regard the present state of the universe as the effect of its anterior state and as the cause of the one which is to follow." In accord with Newtonian mechanics, Laplace elaborated, "Given for one instant an intelligence which could comprehend all the forces by which nature is animated and of the respective situation of the beings who compose it — an intelligence sufficiently vast to submit these data to analysis — it would embrace in the same formula the movements of the greatest bodies of the universe and those of the lightest atom; for it nothing would be uncertain and the future, as the past, would be present to its eyes."

In the Newtonian Cosmos, if the mass, location, and velocity of every particle are given at a particular time, what physicists call "the initial conditions," then Newton's three laws of motion determine the future and the past of the universe. Laplace's super-intelligent being knows the entire history of the universe, from the Big Bang to the Big Freeze, including the rise and fall of the American Empire and me writing and you reading this paragraph.

Laplace's super-intelligent being would be banned from playing roulette, craps, and other games of chance in every Las Vegas casino; for such a being, chance does not exist. The outcome of a ball tossed into a spinning roulette wheel or dice tumbling on the crap table can be calculated with infinite precision. Chance is an illusion rooted in human ignorance.

Psychologist Joshua Greene and neurobiologist Jonathan Cohen give an excellent statement of the determinism adhered to by most scientists: "Intuitively, the idea is that a deterministic universe starts however it starts and then ticks along like clockwork from there. Given a set of prior conditions in the universe and a set of physical laws that completely govern the way the universe evolves, there is only one way that things can actually proceed." In this picture, the current state of the world is completely determined by the laws of physics and by any one of its past states.

Greene and Cohen rightly point out that in the Newtonian Cosmos every event with the possible exception of the Big Bang results from prior mechanical causes. That I would appear in the universe with a small mole on my right temple was in the cards one second after the Big Bang. Yesterday, a long chain of mechanical causes made my wife buy a new Marimekko dress. The precise attunement of particles in the early universe led Francis Crick and James Watson to discover the structure of DNA, Charles Townes to conceive the laser, da Vinci to paint the Mona Lisa, and Mozart to compose the Requiem Mass in D Minor. The deterministic outlook that permeates all science necessarily proclaims that human beings possess no free will, that we are machines, mere pawns moved about by mindlessly forces in a mechanical universe.

The determinism that is a basic aspect of the Newtonian Cosmos is false for two reasons — quantum physics and chaos theory. In the twentieth century, physicists discovered individual events on the atomic level are not predictable. Consider uranium 238, a commonly occurring radioactive substance. A gram of uranium 238 contains approximately 2.56×10^{21} identical uranium nuclei; about 12,300 of those nuclei decay every second into an alpha particle and a thorium 234 nucleus. Quantum physics can predict the probability that a given uranium 238 nucleus will decay but not when it will. The exact moment of decay is intrinsically unknowable. Unpredictability thus is an integral part of nature.

On the macroscopic scale, chaos theory killed determinism. The basic element of chaos theory first appeared in the astronomical investigations of French mathematician Henri Poincaré at the end of the nineteenth century, but only in the twentieth century with the advent of computers did physicists and mathematicians see the full significance of his work. While investigating a system of three gravitating bodies, such as the Earth, Moon, and Sun, the so-called "three-body problem," Poincaré discovered that unpredictable behavior occurs in deterministic systems, a fact that has startling implications for the mathematical modeling of physical systems. Before his revolutionary work, physicists and mathematicians assumed that small errors in the initial conditions of any dynamical system produced only small errors in the mathematical prediction of the future state of the system. Poincaré's analysis of the three-body problem is brilliant and highly technical; yet, the basic result can be easily stated, as Poincaré himself did in his popular book Science and Method: "It may happen that small differences in the initial conditions [of a mechanical system] produce very great ones in the final phenomena. A small error in the former will produce an enormous error in the latter. Prediction becomes impossible, and we have the fortuitous phenomenon."

Vladimir I. Arnol'd, a Russian mathematician, proved, in 1963, that depending upon the initial conditions, the motion of three gravitating bodies can be either predictable or chaotic. Advances in computer technology have made it possible to begin to directly address the long-term evolution of the solar system. The results of Gerald Jay Sussman and Jack Wisdom suggest that the solar system as a whole is chaotic, making its long-term behavior uncomputable. In particular, their computer models indicate that Pluto has a chaotic orbit and that astronomers cannot predict whether the planet will be on this side of the Sun (relative to the Earth's position) or on the other side ten million years from now. In his book Newton's Clock: Chaos in the Solar System, Ivars Peterson concludes, "Long held up as a model of perfection and the symbol of a predictable mechanical universe, the solar system no longer conforms to the image of a precision machine. Chaos and uncertainty have stealthily invaded the clockwork."

The chaos that Poincaré discovered in the Newtonian Cosmos is often called deterministic chaos to emphasize that in an ideal world with infinite precision and without rounding errors in computations, the future behavior of a mechanical system is entirely determined by its initial conditions. However, the Newtonian Cosmos is a map of the physical world drawn with mathematical lines that are breadthless and points that are dimensionless. Mathematical lines and points are idealizations that approximate the physical world. In the Newtonian map, the shape and location of objects can be specified with infinite precession. However, no perfect circle exists in nature, and the boundary of a physical object is fuzzy, not a sharply-defined concept as in mathematics. In the Newtonian map, the laws of physics are eternal, abstract, and never fully embodied in actual matter. Such a mathematical map often helps a scientist to understand a system or situation but the real world always differs significantly from the map.

The Map Is Not the Territory

Polish-American scientist and philosopher Alfred Korzybski enunciated, in 1931, the general principle *the map is not the territory* that applies to all maps.[33] Here map represents an object and can be a guidebook, a history, a novel, or a science. To give one simple example: Years ago, for three summers in a row, I drove from New Hampshire to New Mexico and often consulted an AAA map and guidebook given to me by a well-meaning friend. The static map, of course, did not indicate unpredictable events such as road construction or detours caused by car accidents.

I stopped one night in Springfield, Missouri, and the guidebook directed me to a motel decorated in a Fifties motif and to the Highlander Restaurant.

My AAA road map and guidebook necessarily contained only a broad sketch of the roads, motels, and restaurants along my route. What the guidebook did not tell me was that inside the restaurant, I would hear Howard, a cool middle-aged, black musician, playing an electric organ and singing love songs from the Fifties, nor did AAA tell me I would see senior citizens slow dancing just as they had during their high school days filled with romance. I was overjoyed to discover that some loves last over fifty years. Nor did the guidebook inform me that I would stumble upon a bartender in the heart of the Bible Belt who made a wicked martini. A tourist goes to see a sight described in a guidebook, while a traveler hopes to encounter the unexpected not given on any map.

In the road map I used, the distance on the map from Springfield to Joplin was less than the distance from Joplin to Tulsa, and the same relationship existed in the actual driving distances in the territory. For a map to be accurate, certain structures in it must be identical to the same structures in the territory. Many markings on a map, however, pertain only to the map. Neither I nor any other driver would read an AAA road map literally and then expect to see the Interstate exchanges painted red and blue. The colors made the map easier to read and were not intended to represent anything in the territory.

We should always keep in mind that every map is limited and that very few of the maps we consult for guidance in life are mathematical. For Instance, John Locke's *Second Treatise on Government* is a map of political life, one that gave direction to the Founding Fathers of America; Locke's map does not include anything about *philia* (friendship) or *agápē* (Christian love), which bound together the Greek polis and the medieval village, respectively.

Unlike painters and sculptors, who never mistake the guidebook for the museum, most biologists, neuroscientists, and physicists confuse the maps drawn by their respective sciences with the territory. One reason scientists take the map for the territory was hinted at by Laplace when he visited Napoleon Bonaparte at his country estate, in 1802. The mathematician entertained the French political leader by relating the nebular hypothesis, Laplace's scientific explanation of how the solar system formed from the rotating atmosphere of the primitive Sun. When Napoleon asked Laplace about the place of the Creator in his mechanical account of the solar system, he replied, "Sire, I have no need of that hypothesis."[34] If the universe is deterministic, then a mathematician or a physicist approximates an all-seeing, super-computing god. A mathematician or a physicist, marooned inside his skull contemplating the Newtonian map, believes that he borders on the super-human, for he too can calculate the future and the past of the universe.

Another reason scientists take the map for the territory is that if the limitation of every map is acknowledged, then we humans necessarily have to accept our scientific theories will always be incomplete. To do this requires humility on our part and an acceptance of limits to science itself. No map or theory can capture the concreteness of real life, and any map of the total cosmos that fashions a story about the cosmos and our place in it must fall short in significant ways.

In the Newtonian map, most, if not all, non-chaotic dynamical systems have counterparts in the territory. The Newtonian map by mathematical design is drawn with infinite precision, and thus future behavior of any mechanical system in this map, even if chaotic, is fully determined. But in the territory, no digital clock records time to an infinite number of decimal points, and every meter stick has fuzzy, ill-defined endpoints; thus, the precise initial conditions of a physical system are a fiction, and chaos is real. Nature is not a gigantic clockwork mechanism; the cosmos is not deterministic. The infinitely precise attunement of particles in the early universe is part of the myth of determinism instilled by the habit of taking the Newtonian map for reality. No arrangement of elementary particles ten seconds after the Big Bang led Francis Crick and James Watson to discover the structure of DNA, Charles Townes to conceive the laser, da Vinci to paint the *Mona Lisa,* and Mozart to compose the *Requiem Mass in D Minor.* No human being is a machine; Crick, Watson, Townes, da Vinci, and Mozart were creative minds exploring nature and human life.

When biologists, neuroscientists, and physicists argue 1) the brain is the mind, 2) the brain is determined, and thus the thoughts that arise from the brain are determined, and therefore 3) free will is an illusion, those scientists are describing the map, not the territory. The real, physical world is not a determined machine. Chaos demands indeterminism, which in turn allows for the presence of chance and mind.

On my part, the principle *the map is not the territory* liberated me from materialism, Marxism, capitalism, nationalism, and every ideology that constrains the human spirit and denies that one glory of human life is the spontaneous discovery of the unexpected. In the territory, we should be intellectual travelers, not ideological tourists. We humans are mere mortals, incapable of understanding the totality of being, although virtually no physicist, neuroscientist, biologist — or philosopher or political thinker, for that matter — can resist the temptation to be God, to claim he or she captured the whole show in three equations or in one intellectual insight or in a clever political slogan. We must rely upon multiple maps to guide us in our pilgrimage through life. Plato and Aristotle, Augustine and Aquinas, Sophocles and Shakespeare, Machiavelli and Marx, Newton and Heisenberg, every profound mapmaker reveals a partial glimpse of the territory, yet profound mysteries always remain.

Freedom: A Prerequisite for Science

To convince ordinary people and fallen away believers in materialism like me that "the combined effects of genes and environment determine all [their] actions," Greene and Cohen hope in the future to possess "extremely high-resolution scanners that can simultaneously track the neural activity and connectivity of every neuron in a human brain, along with computers and software that can analyze and organize these data."[35] Then a person, such as you or I, could watch a film of his or her brain choosing between soup and salad. The analysis software would highlight the "neurons pushing for soup in red and the neurons pushing for salad in blue."[36] The film would show "the tipping-point moment at which the blue neurons in [the] prefrontal cortex out-fire the red neurons, seizing control of [the] pre-motor cortex and causing [the person] to say, 'I will have the salad, please.'"[37]

Greene and Cohen imagine this sort of brainware in every future classroom, so "people may grow up completely used to the idea that every decision is a thoroughly mechanical process, the outcome of which is completely determined by the results of prior mechanical processes."[38] (See Figure 2.1.)

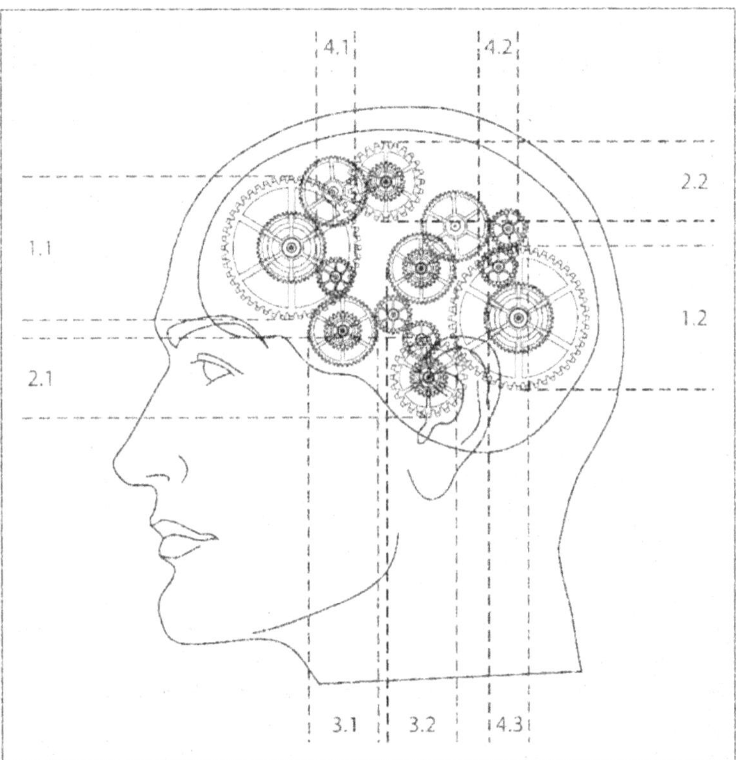

Figure 2.1 An artist's rendition of "every decision is a thoroughly mechanical process, the outcome of which is completely determined by the results of prior mechanical processes." Robert Voight, *Gear of the Human Mind*, Shutterstock.

Suppose such imaginary equipment that Greene and Cohen fantasize about had been available to record the most famous conversations in modern physics, Niels Bohr trying to convince Albert Einstein that quantum physics is true and the founder of relativity attempting the opposite. Using a high-resolution brain scanner, we watch the neurons pushing for the truth of quantum physics in red and the neurons pushing for its falsity in blue. The heated debate ends with two tipping-moments, the red neurons in Bohr's brain out-firing the blue and the reverse in Einstein's brain. Mechanical processes cause Bohr to mumble, "Nature is indeterminate," and Einstein to shout, "No! God does not play dice!" In this scenario, the combined effects of their genes and environments determine the two great physicists' actions; Bohr and Einstein were simply mouthpieces for their differing genes, say numbers 26 and 27 on chromosome 12, and the cultures of their youth, Danish and German.

We have arrived at the Great Contradiction of modern science. If scientists are, as their science proclaims, nothing but a pack of neurons, and their joys, sorrows, memories, and sense of free will are in fact no more than the behavior of a vast assembly of nerve cells[39], then intellectual insight is an illusion. If scientists' thoughts are determined by neurophysiology, action potentials, and the endocrinology of neurotransmitters, then physics, biology, and psychology are meaningless. If every decision in science, as in ordinary life, is a "thoroughly mechanical process . . . determined by the results of prior mechanical processes,"[40] truth is an illusion.

Science, in the twenty-first century, fired a bullet into its own brain.

Let me restate the logical contradiction of the deterministic outlook of science in the simplest possible terms. Suppose my interior life is determined wholly by the motions of atoms in my brain. One day, the atoms in my brain jostle around, and what arises is the ancient Irish belief that leprechauns store their gold in a pot at the end of the rainbow. Such a belief clearly pertains to the atoms in my brain, not to an objective world. But this is true for all my beliefs, and hence I have no reason to suppose that any of them are true, including the one that the motions of atoms in my brain cause my interior life. From this contradiction, we can safely conclude that mind cannot be a mere byproduct of matter.[41]

The incompatibility of determinism and science has been masked over by a sleight-of-hand trick. As we noted in Chapter 1, Galileo, in 1623, argued that tastes, odors, and colors reside only in human consciousness, and all these qualities would be wiped away and annihilated "if the living creature were removed."[42] Since then, science aimed to construct a map of the universe *as if the scientist were not part of it*. The physicist, the chemist, and the biologist in their imaginations watched the universe from the outside, as if from God's former vantage point. Through this stratagem, determinism gives the appearance of being true. For example, when philosopher Bertrand Russell wrote that he looked forward to a "mathematics of human behavior as precise as the mathematics of machines,"[43] he clearly excluded himself, for he set forth his fundamental beliefs in the essay "A Free Man's Worship." Greene and Cohen, too, excluded themselves when they enunciated the fundamental principle of determinism: "Every [human] decision is a thoroughly mechanical process, the outcome of which is completely determined by the results of prior mechanical processes."[44] You and I are mindless machines, but since freedom is a prerequisite for science, Greene and Cohen are not. You and I are "nothing but a pack of neurons"[45] and our sense of free will is an illusion, but in his research program, Crick and his colleagues are not.

In every physics colloquium, I attended in my youth when there were virtually no female physicists, a researcher at the front of the room attempted to *persuade* his colleagues seated in front of him that his theoretical ideas illuminated nature or his experimental results were free from error. In the colloquium room where truth was pursued in public, an unstated premise was that each attendee had the freedom to judge whether something is true or false.

For science to be possible the human will must be free. A denial of free will renders the whole of science absurd. Physicist Carl von Weizsacker states unambiguously, "Freedom is a prerequisite of the experiment. Only where my action and thought are not determined by circumstances, urges, or customs but by my free choice can I make experiments."[46] Cosmologist George Ellis argues that free will is a requirement for any theoretical science to exist: "The enterprise of science itself does not make sense if our minds cannot rationally choose between alternative theories on the basis of the available data, which is indeed the situation if one takes seriously the bottom-up mechanistic view that the mind simply dances to the commands of its constituent electrons and protons, algorithmically following the imperatives of Maxwell's equations and quantum physics. A reasoning mind able to make rational choices is a prerequisite for the academic subject of physics to exist."[47] Neuroscientist John Eccles concurs: "There are thus no sound scientific grounds for denying the freedom of the will, which ironically, must be assumed if we are to act as scientific investigators."[48]

What holds for the scientist also applies to the layperson: Every human has the capacity to make free choices. Genes and the environment do not determine all our decisions. Two core human experiences are not illusions: We can choose one course of action over another, and we can apprehend the truth. Certainly, we often act from mindless habit, and we often accept unexamined opinion as true; nevertheless, we have the capacity to grasp the truth and to choose good over evil.

Every scientist I have met inhabits two worlds, the world of ideas, where humans are machines devoid of free will, and the world of social relations, where people are morally accountable for their actions. For instance, Marvin Minsky, a computer scientist and a leading theorist of artificial intelligence, holds that "according to the modern scientific view, there is simply no room at all for 'freedom of the human will'," since everything in the universe, including what happens in human brains, depends only on "fixed, deterministic laws" and "purely random" events. Yet, Minsky argues that we cannot give up the "myth of voluntary choice [since] our social lives depend upon the notion of *responsibility*." In addition, "we make use of the idea of freedom of will to justify our judgments about good and evil." Minsky concludes that each one of us must accept and also reject the belief that we can voluntarily choose: "We're virtually forced to maintain that belief [voluntary choice], even though we know it's false."[49]

Biologist Richard Dawkins concurs: "Any crime, however heinous, is in principle to be blamed on antecedent conditions acting through the accused's physiology, heredity, and environment.[50] We humans cannot accept such a position "because mental constructs like blame and responsibility, indeed evil and good, are built into our brains by millennia of Darwinian evolution."[51] That each one of us is an intentional agent is a useful fiction programmed into us by natural selection, and that explains why no one — neither scientist nor layman — can live without the illusion of free will. In Dawkins' view, a jury trial to determine guilt or innocence makes as much sense as a man beating his car with a tree branch because it refuses to run.

Presumably, Dawkins would agree with the San Francisco jury that acquitted the former Supervisor Dan White of murdering Mayor George Moscone and Supervisor Harvey Milk. White did not deny his guilt, only culpability. His lawyers invoked the Twinkie Defense. A health-food fanatic, White had gone off his rigid diet and consumed large amounts of Twinkies and Coca-Colas. A psychiatrist at the trial testified that high-blood sugar caused White to "explode" and at the time of the killings he was on "automatic pilot."

Here is an amusing example of the contradictions between the two worlds that scientists inhabit. In my office at Los Alamos National Laboratory, to an audience of one, Roger Peterson, a neuroscientist, repeatedly expounded that natural selection determines the behavior of present-day men and women. Both sexes strive to maximize the number of their genes that will be in future populations. The male of the human species, on the one hand, follows the strategy of copulating with as many mates as possible, in layman's language, "love 'em and leave 'em." Thus, every male *Homo sapiens* has inherited an infidelity gene. The female of the human species, on the other hand, bears children and of biological necessity invests much time and energy in offspring. Consequently, every woman desires a monogamous relationship with a man of high status who can supply food and physical security for her children. Through natural selection, women have a fidelity gene. The opposition of the male infidelity gene to the female fidelity gene gives rise to the universal battle of the sexes. Oddly, Roger had no children but did not declare himself an evolutionary failure.

One afternoon, Roger walked into my office in a rage, swearing a blue streak; his wife, Susan, had run off to Berkeley with a chemist — a mere chemist! — not a physicist or a mathematician or some other elite member of the scientific community. "I can't believe *she* did this to me," the denier of free choice shouted. "The blankety-blank bitch. How could she *choose* him over me!"

A week later, when Susan returned to Los Alamos to pick up her grand-mother's heirlooms, Roger confronted her. She simply told him, "We're all hardwired! Don't blame me! I'm wired wrong!" Roger shouted back, "Don't give me that smart-aleck bullshit! *You* betrayed me!"

Explanatory gaps in science, such as neuroscience not explaining con-sciousness solely in terms of brain function, can be tolerated; scientific hope, even if blind, can keep impossible projects on life support, seemingly forever.[52] Logical contradiction, however, destroys any theory; for it cannot be quaran-tined to a tiny domain. If we admit that A and not A are simultaneously true, then the abandonment of logical contradiction bores wormholes throughout the entire theoretical structure — B and not B can also be true, for any B — and the theoretical structure collapses. Such a logical disaster is traditionally called *ex falso quodlibet* (from the false anything follows). Everything is equally true and false at the same time.

If we are free as doing science demands *and* not free as science proclaims, then every decision is the result of mechanical processes *and* every decision is not the result of mechanical processes; then materialism is true *and* material-ism is not true; then everything arises from matter *and* not everything arises from matter. The determinism of science collapses into the worst kind of the-oretical failure, a jumble of contradictory nonsense. Consequently, the grand pronouncements of science — "the human race is just a chemical scum on a moderate-sized planet;"[53] *Homo sapiens* is "a more-or-less farcical outcome of a chain of accidents;"[54] "human beings are lumbering robots manipulated by genes;"[55] "evil and good, are built into our brains by millennia of Darwinian evolution;"[56] and, "when we die, we die and that is the end of us"[57] — have no credibility.

Only a young child, an inveterate drunk, or a one-eyed scientist can be-lieve in a science that simultaneously declares that a person is free *and* not free, that truth is possible *and* not possible. The great hope of modern science to prove through the experimental method that matter is the ultimate source of everything terminated in nonsense.

3 How Democracy Instills Reductionism

We Americans so firmly believe that each one of us has freely chosen our own way of life that we immediately reject the suggestion that to a large extent our thinking is programmed by culture; I know I did for years. But culturally-given habits and attitudes are especially powerful because we do not usually reflect on them. "Most of us are prisoners of the culture's current presuppositions about life's purposes," historian Richard Rapson observes. "We believe ourselves to be the shapers of our own destinies, but more often we chase after culturally-defined goals as though we were automatons, unaware of the spate of signals which constantly barrage us and mold our attitudes. Even if those goals are achieved, the mature adult frequently finds them unsatisfying and wonders where his or her life went wrong."[58] So, perhaps, just perhaps, we are trapped within the bubble of Modernity and not as free as we think we are.

Every culture tells its members who they are, why they are here, and what the world is about. What defines a culture are the social, emotional, and intellectual habits passed on from one generation to the next.[59] Culturally-given habits are the substance of custom, or "our way of doing things," and they possess more authority than civil law because such habits are the essence of daily living. Every culture lays down patterns of behavior and thought that most of its members follow blindly. Often, instead of a person saying "I am thinking," it would be more accurate for him or her to say that "culture thinks for me." Psychoanalyst Erich Fromm elaborates: "In expressing an opinion, for instance, we say, 'I think' this or that. If one analyzes this opinion, however, one might discover that the person only voices what he has heard from someone else, what he has read in the newspaper, what he was taught by his parents when he was a child. He is under the illusion that it is he who thinks of this, when actually it would be more correct if he said: 'It thinks me.' He has about the same illusion a record player would have which, provided it could think, would say, 'I am now playing a Mozart symphony,' when we all know that we put the record on the record player and that it is only reproducing what is fed into it."[60]

Ethnologist S. M. Molema, writing about his own people, the Bantu tribe, points out that in premodern Africa, a person's "actions are controlled by iron reins of tradition, his conduct is constrained by rigid custom. *His very words are often a formula.*"[61] Prince Modupe of the So-so tribe confirms Molema's observation: "When I lived with my people in Dubricka, all of my opinions and judgments were formed by them, by my mother and the elders and the teachers in the Bondo Bush. A youth was taught never to question the validity of anything an elder said."[62] Many Eskimo and aboriginal tribes do not even have a word for disobedience. Group expectations are so strong that the young learn to follow custom unthinkingly, and as a result, such ancient peoples have no system of law and punishment.

Ancient Habits of Thinking

A person in a premodern culture acquires at an early age the habit of understanding himself as part of a larger whole. Tocqueville emphasizes that an "aristocracy links everybody, from the peasant to the king, in one long chain."[63] If I am a link in such a chain, I understand myself by looking outward to see who is above and below me in the social hierarchy. In this way, I form the habit of understanding myself as part of a whole. Later, I transfer this habit to thinking about other things. Thus, in a group-centered culture, the first habit of thinking is: *To understand something, see how it is related to the whole.* With such a habit, a person approaches every problem as an organic whole, not a composite whole made up of a sum of parts, and understands each of its many interrelated parts in terms of that whole, whatever it may be — humanity, the family, the person, or even an animal or plant. Before a Hopi potter begins to shape the clay, she has the entire design of the pot formed in her mind. No single element of the design has a symbolic significance in and of itself, but only in relation to the whole. The design intricacies of Hopi pottery are shown in Figure 3.1.

In the ancient world, the constant reference point is the group. To be separated from the group is to lose one's identity or even one's existence. Modupe says that at the turn of the century in Africa, "Any destiny apart from the tribe was, of course, beyond the limits of either imagination or intuition. It was as unthinkable as that one of the bright orange legs of a millipede should detach itself from the long black body of the creature and go walking off by itself."[64] If I am a member of a group-centered culture, I believe that I am social by nature and that without the group, I would not exist. Each person about me understands himself or herself as part of a group. I see that I am always in need of other persons; thus, when I do not know something, I seek out a person who possesses knowledge and wisdom. In this way, I form a second habit of thinking: *Seek guidance from masters.* Hopi children frequently hear from their parents, "Your old uncle taught us that way; it is the right way" and "Listen to the old people; they are wise."

I also realize my experience is neither unique nor private; I understand what happens to me in terms of experience common to my family, to my clan, or to humanity. I recognize that my understanding is limited, but I have a common treasure to draw upon — the accumulated knowledge of my people.

Figure 3.1. Nampeyo, Hopi ceramic pot, circa 1880. Courtesy Phyzome, Wikimedia Commons

Life, for me, is governed less by abstract thinking and more by a common store of wisdom. Thus, I acquire a third habit of thinking: *Experience will confirm the truth of what the masters say and reveal the wisdom behind their words.* Hopis say, "Our way of life was given to us when time began."

The habits of thinking of group-centered peoples do differ in some important ways. For instance, the Eastern Indians emphasize universals, while the Chinese concentrate on particulars.

Modern Habits of Thinking

Tocqueville, in "Concerning the Philosophical Approach of the Americans," an absolutely brilliant chapter of *Democracy in America*, argues that since an American always begins with the self, each citizen forms the intellectual habit of looking to the part, not to the whole, and as a result is a Cartesian reductionist: "Of all the countries in the world, America is the one in which the precepts of Descartes are least studied and best followed."[65] Tocqueville explains this paradox. In a modern democratic society, the links between generations are broken, and consequently, in such a society, men and women cannot base their beliefs on tradition or class. Social equality produces a "general distaste for accepting any man's word as proof of anything."[66] Therefore, "in most mental operations each American relies on individual effort and judgment."[67] Just like Descartes, each American employs the philosophical method to seek for the reason of things for oneself and in oneself alone.[68]

One general conclusion Tocqueville draws from his study of American life is that "the Americans have needed no books to teach them philosophic method, having found it in themselves. Much the same can be said of what has happened in Europe."[69] Francis Bacon, in natural science, and René Descartes, in philosophy, "abolished accepted formulas, destroyed the dominion of tradition, and upset the authority of the masters."[70] Luther, Voltaire, and several centuries later, the man on the street in America submitted traditional beliefs to individual examination.

Proceeding by leaps and bounds, Tocqueville does not stop to give in detail the modern habits of thinking, so at the risk of appearing slightly redundant, let me flesh out his insights.

First, let me note that to grasp the habits of thinking of premodern peoples, I had to imaginatively use published accounts by Africans, Native Americans, and Chinese, while to understand modern habits of thinking, I just had to look at myself.

As a member of a modern democratic culture, I could not base my beliefs on tradition, custom, or class, because the links between "the peasant and the king" no longer exist in Modernity. By the sixth grade, I did not understand myself in terms of either family or nature; my Romanian heritage meant nothing in America, and although I spend my boyhood summers playing in rural Michigan, I never received any instruction at home or in school about my connection to nature. I understand myself as an isolated, autonomous individual. My constant reference point, then, was always myself. Consequently, I formed the habit of always thinking of myself in isolation from other persons, and this habit carried over when I thought about other things. Thus, my first culturally-given habit of thinking was *To understand something isolate it, so it exists apart from all relations.*

Hence, I believed that every part can be separated from the whole and that the whole can be understood as simply a collection of parts. With such a habit of mind, I attempted to understand every whole solely in terms of its parts. But the smallest parts of anything are material. Consequently, the culturally-given habit of thinking the whole is a collection of parts made me a firm believer in materialism — I could not think any other way. I just "knew" that the universe, including all aspects of human life, was the result of the interactions of little bits of matter.

When I was a young theoretical physicist, I would have staked my life on the proposition that matter is the ultimate reality. The philosophers and aspiring poets I knew in graduate school often asked me over beer and pizza about the fundamental elements of reality. With no hesitancy, like Eric Fromm's record player, I could not help but say, "atoms, genes, individuals, competition, and warfare," and, yet, believed I was thinking, not mindlessly repeating what had been programmed into me, and my philosopher and poet friends did not strenuously disagree. I now know that virtually every American forms the intellectual habit of looking to the part, not to the whole, and thus is at heart a Cartesian reductionist.

Because of the principles of social equality that I had taken in, I did not trust the authority of any person and had an intense "distaste for accepting any man's word as proof of anything."[71] As a result, in most mental operations, I relied on my own judgment and thought. Although I found philosophy a bore and totally irrelevant to my life, yet, I proceeded just as Descartes did; the intellectual method I employed was to seek by myself and in myself "for the only reason for things."[72] Since American culture told me that all individuals are equal and that I could recognize the truth just as well as the next person, I thought that I had no need to seek guidance from others, even acknowledged masters. Indeed, I believed that if I followed another person's judgment, I would give myself over to that individual and thereby enslave myself and violate what was most precious to me, my personal freedom. Thus, my second habit of thinking was *Rely solely on individual judgment and thought.* Consequently, in American life no masters are recognized, and, in effect, the three great teachers of humankind — the Buddha, Socrates, and Jesus — are just three voices among many. In fact, if anyone holds up someone as a master to follow, most Americans will intentionally ignore or dismiss that person, since it smacks of inequality.

American culture also informed me that the essence of individuality is uniqueness. Each individual has his or her own unique beliefs, tastes, feelings, thoughts, desires, and expectations. What is true for another individual is not true for me: Everybody is different. Furthermore, each individual has a different way of evaluating his or her experiences; another individual's word or experience is not proof of anything. However, since all individuals are equal, my direct experience is not proof of anything either. From these cultural opinions, I learned to distrust my own experience. For instance, when I read in Aristotle's treatises that "the whole is prior to and greater than the part," "every person desires to know," and "man is social by nature" are first principles, I did not look to my own experience for confirmation, but instead demanded a proof of some kind, and not finding an acceptable proof, I took each of these statements as assumptions that I could later deny if I so wished.

Since individual experience is unique and truth is universal, I was led to form a third habit of thinking: *Accept as true only what can be proved through logic, mathematics, or scientific experiment.* For me, reason and the scientific method replaced the authority of direct, concrete experience. In principle, I could carry out any logical argument, mathematical demonstration, or scientific experiment; thus, I never had to submit myself to an acknowledged master or any outside authority. I readily accepted the results of electrodynamics, general relativity, and quantum mechanics because they made no claim on my interior life and never challenged who I took myself to be.

Layered on top of these three democratic habits of thinking are religious dogmas and political ideologies — the equality of conditions does not exist in a vacuum. Many political junkies, those rabid viewers of Fox News or MS-NBC, *believe* that they rely solely on their own judgment; when in actuality, political ideologues, left and right, often parrot what they saw on TV, heard on talk radio, or read on the Internet. Religious dogma and political ideology led to creationism and to the denial of climate change, despite scientific evidence.

No one doubts that many Americans today consult priests or psychotherapists for guidance in life, although I suspect not with the blind faith or trust they would have had in the Fifties. With the intensification of equality in the Sixties, parishioners and clients are the final judge of what is best for them.

That science is the only intellectual authority in Modernity seems to be contradicted by the widespread belief in astrology, herbal cures for cancer, and crystal-healing of a disturbed psyche. But when science proclaims life is pointless, can offer only slash-burn-and-poison for cancer treatment, and restores mental well-being through tranquilizers and psychotropics, then even the college-educated out of desperation turn to alternatives.

The real alternative is to examine the modern habits of thinking, all of which rely upon the links of the long chain from "peasant to king" having been broken. Said in a more general way, in contrast to premodern cultures where people understand themselves to exist only in relationship, in Modernity, individuals assume they exist in isolation. One of these understandings of the human being, clearly, must be wrong.[73]

Nihilism: Another Pathology of Modern Life

In examining the habits of modern thinking, I accidentally uncovered another intellectual pathology of our day, one I subsequently called American Nihilism, for it stems directly from democracy.

All the scientists I know recognize the limitation of science to give meaning to their lives. Last week at lunch, a physicist friend of mine, a former staff member of Los Alamos National Laboratory and the owner of a consulting company in Santa Fe, told me, over glasses of Johnny Walker Black, there are "two kinds of truth, one scientific, the other personal." He, then, added, "Most of us just want to believe what makes us comfortable and not have our beliefs challenged, and I am okay with this; I do it myself." When I pressed him, I hope tactfully, but now I am not so sure, he said that scientific meant universal truths and that personal truths are subjective, unique to each individual, but necessary to live. I pointed out "personal truth" is modern cliché, a misuse of the word "truth," a euphemism for "personal opinion;" he just shrugged his shoulders and said, "That is what I believe."

All my physicist friends hold that material things can be known through and through, while beauty, human values, and the purpose of human life are unverifiable opinions, true for the individual who holds them, but not necessarily for the rest of humanity. My former colleagues readily accept scientific truths, say the results of electrodynamics, general relativity, and quantum mechanics because these disciplines do not challenge how they live. Every time I asked them to articulate their personal truths, I heard some combination of the usual American goals: be a success, amass wealth, and acquire a house with a white picket fence, a dutiful brunette wife, and a golden retriever named "Rex." When I pressed them to examine their personal truths, they refused either from a willingness to rest contentedly in ignorance, as Socrates would hold, or from the fear of falling into the abyss of nihilism, as Nietzsche would maintain.

Another friend of mine, Judge Gary Carlson, presided for years over a juvenile court in a small town in Illinois. He repeatedly laments that he and the young offenders who appeared before his court shared no common "moral core." When Gary from the bench advised a young person how to change his life for the better, he invariably heard, "Judge that is your opinion, not mine. Heh, man, don't you know, it's different strokes for different folks."

Shockingly, the opinion of a semi-literate juvenile in rural Illinois is fully articulated in a United States Supreme Court Ruling written by Justice Anthony Kennedy: "At the heart of freedom is the right to define one's own concept of existence, of meaning, of the universe, and of the mystery of life,"[74] a ruling in keeping with the widespread opinion that everyone has his or her own personal belief system — even the right to one's own concept of the universe!

The danger inherent in American nihilism is that democracy will become a demagoguery. With tradition a mere curiosity touched upon in grade school, with no common core of beliefs other than equality and individualism, public discourse cannot be moderated and directed by reason. In the marketplace of ideas, rational argumentation disappears, replaced by passions and prejudices, a phenomenon most evident on the internet. Anyone, anywhere, anytime, now, can instantly post an opinion on anything. Not constrained by historical facts or scientific truths, the postings on the internet constitute an ocean of opinion, one comment washing over another, quickly submerging whatever truth tries to surface, in effect, generating the Age of Post-Truth. Political discourse becomes personal, emotional, and opposing viewpoints irresolvable. My passions are my truths; I scoff at the truth that I am a leaf that is part of a tree. To me the best political leader mirrors my feelings and prejudices, say rage at job loss, disappointment that the church, the union hall, and the VFW are not the places of camaraderie they used to be, and hatred of immigrants destroying the American way of life, or perhaps contentment with economic success, pleasure of breaking free of family and social restraints, and joy over the Supreme Court legitimizing same-sex marriage.

As we pointed out in the Introduction, Friedrich Nietzsche was the first philosopher to see the undercurrent of nihilism running beneath Modernity. In his 1873, unpublished essay, "On Truth and Lie in an Extra-Moral Sense," he asks, "What is truth?" and answers, "A mobile army of metaphors, metonyms, and anthropomorphisms."[75] One hundred years later, in an essay published by the American Council of Learned Societies, six eminent professors of literature assert, "All thought inevitably derives from particular standpoints, perspectives, and interests."[76]

Unlike Nietzsche, who lived on the edge of despair because truth could not be known, we moderns secretly rejoice over the "advent of nihilism."[77] In the absence of truth, we are totally free, for we do not have to submit to anything beyond ourselves. Each one of us is the sole judge of what is true, good, and beautiful; as "King of the Castle," we are demigods, inventing ourselves, devising our own ends, and accepting or rejecting whatever we wish. Writer David Foster Wallace, in his Kenyon College Commencement Address, in 2005, warned the graduates, "Our own present culture has harnessed these forces ['fear and contempt and frustration and craving and the worship of self'] in ways that have yielded extraordinary wealth and comfort and personal freedom. The freedom to be lords of our own tiny skull-sized kingdoms, alone at the center of all creation."[78]

Tocqueville never foresaw the danger freedom posed to American democracy. When he visited America in the early days of the Republic, he observed that unlike Europe the spirit of religion and the spirit of freedom formed a "marvelous combination,"[79] a harmonious balance. For him, the "founders of New England were both ardent sectarians and fanatical innovators." He admired how religion and freedom were companions in the struggle to establish democracy in the New World: "In the moral world, everything is classified, coordinated, foreseen, and decided in advance. In the world of politics, everything is in turmoil, contested, and uncertain. In the one case obedience is passive, though voluntary; in the other, there is independence, contempt of experience, and jealousy of all authority." Tocqueville failed to see that the spirit of freedom was unstoppable and could not be contained to politics; eventually religion would be eroded by the "jealousy of all authority."

From the standpoint of the Western intellectual tradition anchored in Athens, Americans have embraced a mistaken notion of freedom, one that I, too, embraced in my youth and still held when at Los Alamos Scientific Laboratory. For Socrates, Plato, and Aristotle, truth is primary, not freedom, and the inherent danger in human life is to forsake reason and become a slave of the passions. The good life consists in the "active exercise of the soul's faculties in conformity to rational principle,"[80] which is Aristotle's way of saying that for a person to be happy the passions must be directed by reason. Courage frees a person from slavery to fear, generosity from the slavery to hunger for money; temperance from slavery to drugs, alcohol, and sexual lust. Without freedom, a person could not acquire good habits or replace bad habits with good ones, and thus would be condemned to a life of slavery. Self-mastery is the mark of a free person, not the "license to do whatever one wants."[81]

According to ancient wisdom, the only way out of American Nihilism is to make truth primary and to adopt the mantra that freedom is obedience to truth. Unlike nihilism, reductionism is a pernicious, difficult disease to treat.

4 Dumb Idea # 2: Reductionism

In January 2011, an intriguing announcement arrived in my email inbox. *The New Yorker*'s upcoming issue was to contain "Social Animal" by David Brooks, *The New York Times* columnist and guru of middle-class American life. I could hardly wait to read "how the new sciences of human nature can help make sense of a life" like mine and those other lost souls around me in Santa Fe, New Mexico.

Three days later, my mailwoman, Kate Bowman, delivered to my front door *The New Yorker* I eagerly awaited. Kate had jettisoned her Harvard Ph.D. in philosophy for the spiritual landscape of Northern New Mexico. I asked her what she thought of Brooks' article. She moaned, rolled her eyes, and said, "Now, I know that I was right to leave academia — unadulterated horseshit is fobbed off as pearls of wisdom." I always suspected Kate had become disillusioned in graduate school, and consequently, I ignored her parting shot: "See if you think social scientists have made great strides in understanding human nature, filling in the atrophy of theology and philosophy." Another moan, and then, "Give me Adi Shankara any day." I shrugged off the mention of a ninth-century AD Hindu guru, for in Santa Fe even my furnace repairman teaches Sanskrit.

Speed-reading through the "Social Animal" in desperate search for profound truths about human life, I, too, moaned and rolled my eyes as I read about how the two made-up characters, Harold and Erica, sized each other up when they met for the first time. "Harold liked what he saw, from the waist-to-hip ratio to the clear skin, all indicative of good health and fertility."[82] Erica's smile pleased him, and he "unconsciously noted that the end of her eyebrows dipped down." Brooks reported that Duchenne Guillaume, a French physician, observed in the mid-nineteenth century that the outer portion of *orbicularis oculi* muscle cannot be voluntarily controlled, so a smile where the tip of the eyebrow dips betrays a spontaneous, genuine emotion.

Erica was impressed by Harold's symmetrical features, a sign of a strong immune system, and that he was slightly older, taller, and stronger than she, indicative of a good provider. Deep down in her genes, she knew that "while Pleistocene men could pick their mates on the basis of fertility cues discernible at a glance, Pleistocene women faced a more vexing problem." Since a Pleistocene woman wrapped in a bearskin could not supply sufficient calories for a human baby that required years of care, she had to choose a mate "not only for insemination but for continued support." Those Pleistocene women who chose short, weak, or old men with unsymmetrical features bred successfully, but their offspring perished; in this way, natural selection gave present-day women the genes to select successful mates.

To test the "science behind everyday life," I went to the local Whole Foods Market to collect my own data. The tape measure never left my pocket, for I immediately saw that not one woman in the upscale grocery store exhibited the hourglass figure, a waist-to-hip ratio (WHR) of .7 that Marilynn Monroe and Sophia Loren had, and supposedly desired by all men, according to Brooks. I suspected that buried in the scientific literature was a paper that purported to show that many indigenous peoples, such as those in southeast Peru, who have had little contact with the Western world, prefer high WHRs. These researchers, no doubt, concluded that a penchant for low WHRs is a byproduct of modern Western culture. Later, I did check the scientific literature and discovered that 88 percent of American women fit into four basic body types: 8 percent have the curvaceous hourglass figure flaunted by 1950s movie starlets; 46 percent are banana-shaped; 20 percent are bottom-heavy pears; and 14 percent broad-shouldered apples. *Vive la difference*!

These days no matter where I turn, I encounter questionable opinions about human nature based on evolutionary psychology and a neuroscience whose goal is to prove all members of *Homo sapiens* are mere machines. Even my roofer — every flat roof in Santa Fe leaks at some time — told me over a cold beer that morality evolved in our hunter-gather ancestors to benefit their selfish genes. No doubt, he, like Brooks, was merely parroting mainstream evolutionary thinkers and their popularizers. Sociobiologist Edward Wilson and philosopher of science Michael Ruse assert that ethics is a biological adaptation to further reproductive ends and that the belief in the Golden Rule is a shared illusion of the human race.[83] That evolutionary psychology speaks of moral behavior caused by genes, not moral choice, is an outlook that stems from reductionism, the assumption that every whole is completely understandable in terms of its smallest parts and how they interact.

Francis Crick, the co-discoverer with James Watson of the double-helix structure of DNA, gives an example of how reductionism sets the agenda of neuroscience: "The aim of science is to explain *all* aspects of the behavior of our brains, including those of musicians, mystics, and mathematicians. To understand the brain, we may need to know the many interactions of nerve cells with each other; in addition, the behavior of each nerve cell may need explanation in terms of ions and molecules of which it is composed. Where does this process end? Fortunately, there is a natural stopping point. This is at the level of chemical atoms."[84]

In the grand scheme of reductionism, each higher level is explained in terms of the next lower level: society by its individuals, an organism — human or otherwise — by its organs, an organ by its cells, a cell by its molecules, a molecule by its protons, neutrons, and electrons. If reductionism is true, then Wilson correctly concludes, "All tangible phenomena, from the birth of stars to the workings of social institutions, are based on material processes that are ultimately reducible, however long and tortuous the sequences, to the laws of physics."[85] According to this outlook, historical events are reducible to sub-atomic events. In principle, the near-collapse of the world economy in 2008 or the election of a new president in the same year can be explained in terms of quarks and leptons. Biologist Peter Medawar, however, finds this nonsensical and ludicrous: "There is simply no sense in saying that politico-sociological concepts like electoral reform and the foreign exchange deficit can be 'interpreted in terms of biology,' and it is hardly less than idiotic to say that they can be interpreted in terms of physics and chemistry, though if the axiom of reducibility were true it would follow that they were so."[86] Is reductionism a reasonable goal of science or an "idiotic" quest?

Organic and Composite Wholes

To answer this question, first consider DNA, thought by many to represent the final victory of reductionism in biology. By analyzing data from x-ray crystallography, Crick and Watson, using only physics and chemistry, established that the physical structure of the DNA molecule is a double helix. But why DNA has the form of a double helix can be answered only by looking to a larger whole, the cell. When a cell divides, the DNA in the cell replicates itself by unzipping into two helical strands and incorporating new molecules from the surrounding medium in precisely the right order to form two identical copies of the original double helix. (See Figure 4.1.) Thus, in molecular biology, the physical structure of a part, DNA, is understood in terms of the whole, the cell.

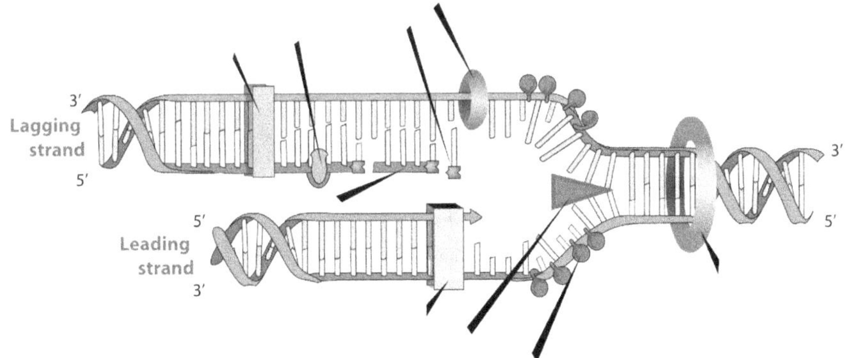

Figure. 4.1. The DNA molecule unzipping into two helical strands (Wikimedia Commons).

The units of heredity are genes. In molecular terms, a gene is simply several stretches of DNA. Like strings of letters taken from the entire alphabet, the chemical sequences that make up DNA have no meaning in themselves. The meaning of a string of letters is determined by a whole; information theorist Hubert Yockey gives an amusing illustration: "*O singe fort!* has no meaning as a sentence in English, although each is an English word, yet in German it means *O sing on!* and in French it means *O strong monkey!*"[87] Similarly, what particular sequences of chemical units have meaning as a gene can only be determined by looking to a whole — to the cell, to the organism, or to the pedigree of the offspring.

Huntington's disease, for example, characterized by jerky, involuntary movements called chorea, and by mood swings, slurred speech, and eventually dementia, was first diagnosed in 1872 by New York physician George Huntington. He realized it ran in families. Geneticists later traced the disease back to two brothers from Suffolk, England, who immigrated to America in 1630. Huntington's disease follows the classic Mendelian pattern of inheritance. Given a family history of the disease, Mendelian genetics can assign a probability of an offspring getting the disease but cannot predict with certainty who will.

Nancy Wexler, a clinical neuropsychologist, had a fifty percent chance of developing Huntington's disease since her mother died of it. Determined to find the Huntington gene, she traveled in 1979 to three small villages in Venezuela on the shores of Lake Maracaibo, where Huntington's disease is common. The disease had been traced to Maria Concepción, who lived in the early 1800s. She had 11,000 descendants, 9,000 of whom were alive during Wexler's hunt for the gene. Wexler and her team sent blood collected from 500 people to James Gusella at Massachusetts General Hospital in Boston. He searched for random genetic markers. In 1983, when he found one, he narrowed the gene's location to the tip of the short arm of chromosome 4.

Ten years later, six research teams, comprising 52 investigators from ten institutions, announced that the Huntington gene had been found. The gene responsible for the havoc wrought by Huntington's disease is carried by all human beings; however, individuals with the disease have a defective version. One end of the defective gene repeats too often, what is in effect the molecular equivalent of a stutter: Unaffected people have fewer than 35 repeats, while people with more than forty are certain to develop the disease as adults; in those rare cases of more than sixty, a severe form of the disease occurs before the age of twenty.

The extensive search required to find the Huntington gene clearly shows that for a physicist or a chemist, a gene is simply a sequence of chemical units, but for a biologist the function of a gene is determined by looking to some whole. No biochemist or geneticist could have looked solely at the chemical structure of a defective Huntington gene and concluded that it was the gene that causes Huntington's disease. No gene is understandable in terms of itself alone.

Contrasting modern biology with Newtonian physics, we see that there are two kinds of wholes, organic and composite, that differ radically in how a part is related to the whole. An open encounter with any organism reveals that each part, in some way, *always* contains the whole. Consequently, a part is not separable from the whole, for if it could be separated, the part would no longer exist. Consider a jackrabbit in the desert Southwest, for example. The DNA in every cell of its body is *unique* to this individual jackrabbit. The entire rabbit is contained in every cell of its body; in a liver cell, for instance, is the information to build a pancreas, an eye, a brain, and every other part of the rabbit. If the whole were removed from a cell, it would be destroyed; no rabbit cell can exist without its DNA. Just as the Huntington's-disease gene could be understood only by looking to a whole; so, too, rabbit DNA can be understood only by looking to the organism. The rabbit cannot be explained solely by its organs, by its cells, or by the molecules in its cells because each of these parts also contains the whole.

Even the environment of the rabbit is present in some way in its parts. The coyote is present in the rabbit's powerful back legs, desert plants in its sharp teeth, the earth's gravitational pull in its bones, an oxygen-rich atmosphere in its lungs. The history of the universe is also present in the rabbit. The calcium atoms in its bones, the iron atoms in its red blood cells, indeed, every chemical element in its body came from stars that exploded billions of years ago. The matter of the rabbit is literally star stuff, and therefore the Big Bang, galaxies, and stars are present in the rabbit. Remove all traces of the atmosphere out of the rabbit, then its ears, lungs, and blood would no longer exist. Remove the earth's gravity from the rabbit, then its bones would vanish. Remove the Big Bang, and nothing exists. The rabbit exists only as a part of a larger whole. If all the rabbit's relations to its world could be eliminated, then it would cease to be.

Reductionism, then, rests on the assumption that *the part is completely separable from the whole and understandable in and of itself*, which clearly is false for living organisms. In the Newtonian map of nature, where the primary constituents are separable pieces of matter moved by mindless forces, plants and animals appear as inanimate objects, no different than falling rocks or flowing rivers. However, a living organism must be understood as a whole, not as a summation of genes.[88] The failure of reductionism in biology does not mark the failure of science, but only the replacement of an unworkable assumption with a new mode of understanding that recognizes that in a living organism the part can only be understood in terms of the whole.

Unlike the organic wholes of biology, the wholes of Newtonian physics are composite in that they can be separated into parts without destroying or altering the part. In Newtonian physics, the motion of a whole body is determined by the motion of its parts. The gravitational attraction of the Sun on each part of the Earth causes the entire Earth to move in an elliptical orbit around the Sun. Planets, clouds, oceans, and manmade objects do not possess the unity that organisms have. In biology, every part contains the whole in some way; a drop of human blood uniquely determines the person it came from. In physics, the part is completely separable from the whole; a lug nut from a 2003 Toyota Land Cruiser does not uniquely determine the individual vehicle it came from. For lifeless matter, the part *is* separable from the whole, and in this very limited region of nature, the principle of reductionism does apply.

The majority of scientists believe that reductionism is an essential element of the scientific method. Theoretical physicist John Polkinghorne, for instance, assumes that "when we begin to consider the nature of physical reality it is *instinctive* to turn to the insights of so-called fundamental science, to start with elementary particle physics."[89] If beginning with the smallest parts were an instinctive or natural inclination of the human mind, we would find it in each human being, irrespective of culture or historical era. And, of course, we do not.

A few scientists, like geneticist Richard Lewontin, understand that the cultural influence that permeates science "comes in the form of basic assumptions of which scientists themselves are usually not aware yet which have profound effect on the forms of explanations."[90] He adds that reductionism — "that the whole is to be understood *only* by taking it into pieces, that the individual bits and pieces, the atoms, molecules, cells, and genes, are the causes of the properties of the whole objects and must be separately studied if [scientists] are to understand complex nature" — is a cultural belief that stems from individualism.[91] As we saw, one of the habits of thinking instilled by modern culture is *to understand something isolate it, so it exists apart from all relations.*

In the territory of nature, in contrast to the Newtonian map of the cosmos, living organisms are not mere collections of cells, nor "essentially bundles of simple quarks and electrons," as physicist Murray Gell-Mann maintains.[92] You and I are not the sum of our brains, hearts, livers, glands, and nerves, nor are the various communities we belong to a mere collection of isolated individuals. To say, as Crick does, that we are "nothing but a pack of neurons"[93] is clearly "idiotic."

Since the source of reductionism is not in the philosophy of Descartes or in science but in democratic culture, we must turn to a brief examination of how individualism arose to dominate Western culture.

5 Individualism: The Heart of Modernity

While traveling through the dense woods in Michigan, in 1831, Alexis de Tocqueville came across a pioneer and his family, making the "first step toward civilization in the wilds."[94] He noted in his travel diary that "from time to time along the road one comes to new clearings. As all these settlements are exactly like one another, whether they are in the depths of Michigan or just close to New York, I will try and describe them here once and for all."[95] The settler's log house showed "every sign of recent and hasty work." The walls and roof were fashioned from rough tree trunks; moss and earth had been rammed between the logs to keep out cold and rain from inside the house. The settler exhibited little curiosity in his French visitor, and in showing hospitality to the stranger, he "seemed to be submitting to a tiresome necessity of his lot and in it saw a duty imposed by his position, and not a pleasure." The pioneer and his family formed a "little world" of their own, an "ark of civilization lost in a sea of leaves. A hundred paces away the everlasting forest spread its shade, and solitude began again."

Individualism allowed for the rapid settlement of America, as seen in the pioneer Tocqueville described. A landed gentry, an established church, and a class based on birth were left behind in Old Europe. Individual freedom in the New World and the desire for material gain and economic independence unleashed the great potential hidden within every person. As a result, America "opened a thousand new roads to fortune and gave any obscure adventurer the chance of wealth and power."[96] Surprisingly, the ill effects of individualism were seen by some of the early settlers of America. "Thousands of Europeans are Indians, and we have no examples of even one of those Aborigines having from choice become Europeans," Hector de Crèvecoeur, a French émigré, complained in 1782. He thought, "There must be in their social bond something singularly captivating and far superior to anything to be boasted among us."[97]

One hundred and twenty-five years after Tocqueville visited the settler in the wilds of Michigan, I, as an adolescent, forty miles west of Motown, inherited the pioneer's individualism through American culture. I thought that to be free meant to be an isolated, autonomous individual and believed that I owned myself and my abilities, that all the relations I had with other persons I voluntarily chose, and that I owed nothing to other persons except what I of my own free will incurred. The pioneer, living in an "ark of civilization lost in a sea of leaves," was more extreme than I was. He had left behind in Old

Europe parents, grandparents, siblings, aunts, uncles, and cousins, everyone to impede his freedom and independence.

In America, individualism was not an idea found in philosophical treatises and then put into practice, but a lived experience. Despite that, the philosopher John Locke developed the logic of the modern democratic state that the Founding Fathers later appealed to. When arguing for the adoption of the proposed United States Constitution, in 1787, James Madison invoked John Locke's conjectures about how political societies formed.[98] In a state of nature, each individual, alone, "free, equal, and independent," is "constantly exposed to the invasions of others;" property is "very unsafe, very insecure;" and existence is "full of fears and continual dangers."[99] For the protection of his goods and life, no individual can rely upon the goodwill of others, consequently, individuals contract with each other to hand over their natural power to protect themselves and their property to the State. For Locke and Madison, self-interest, weakness, and natural enmity caused isolated individuals to form political societies.

Locke also laid the theoretical foundation of capitalism. He believed that God had given the earth and its fruits to mankind in common: "The earth He has given to the children of men."[100] How, then, could any man have a right to private property? Locke begins to answer this question by arguing that in a state of nature each individual has the natural right to life and thus to the fruits of the earth to sustain his life. A man is the "absolute lord of his own person and possessions;"[101] he has a "property in his own person" that "no one has a right to but himself."[102] When a person mixes his labor with anything in the commons, he thereby makes it his property, for his annexed labor makes it part of himself. When a man killed and skinned a deer or picked a basket of blueberries, he mixed his labor with the game or the berries, and they became his.

Notwithstanding the vastness and abundance of the earth — "God has given us all things richly"[103] — nature limits what a man can rightfully take from the commons. A man may appropriate through his labor as much as he can use before it spoils. No rational man would kill ten deer only to have the meat putrefy or pick two hundred pounds of blueberries only to have the fruit spoil in a week. "Nothing was made by God for man to spoil or destroy."[104]

The second stage in the development of private property occurred with the introduction of money, "some lasting thing that men may keep without spoiling."[105] Gold, diamonds, or whatever men by consent deemed valuable allowed a person to legitimately accumulate as "much of these durable things as he pleased,"[106] and not violate the limits of possessions imposed by nature since money was not perishable.

With money, the industrious enlarged their possessions, and because man by nature has an unlimited desire for material goods, soon the commons vanished, replaced by vast estates. "Since gold and silver, being little useful to the life of man, in proportion to food, raiment, and carriage, has its value only from the consent of men … it is plain that the consent of men have agreed to a disproportionate and unequal possession of the earth …"[107] The introduction of money destroyed the freedom and equality of all individuals that existed in the first stage of the state of nature, and governments had to be instituted to safeguard unequal property.

Men without land were forced to sell their labor for wages; the buyer owned their labor, and thus what they produced belonged to him. The buyer and seller of labor are linked by the exchange of money, not in any permanent way by custom or obligation. According to Locke, unlimited private property and the selling and buying of human labor, the two essential components of capitalism, are rooted in the nature of man. The interference with free markets and class structure by a society or a government goes against God and nature.

The Three Historical Events that Launched Modernity

1. The Black Death killed one-quarter to one-third of the population of Europe and thereby greatly weakened community life. Giovanni Boccaccio reports the extreme degree to which social relations deteriorated when the plague struck Florence in 1348: "One citizen avoided another, hardly any neighbor troubled about others, relatives never or hardly ever visited each other. Moreover, such terror was struck into the hearts of men and women by this calamity, that brother abandoned brother, and the uncle his nephew, and the sister her brother, and very often the wife her husband. What is even worse and nearly incredible is that fathers and mothers refused to see and tend their children, as if they had not been theirs."[108]

2. The rise of the mercantile states of Italy generated enormous wealth and focused attention on the good life in this world, away from salvation. The first modern bank, *casa di S. Georgio*, was founded at Genoa in 1407. Soon, other banks, such as the *Centurioni* at Genoa, the *Soranzo* at Venice, and the *Medici* at Florence, carried on trade in both money and merchandise. Thus, in the Italian city-states the emphasis in daily living shifted from spirituality to worldly gain, personal ambition, and self-interest.

3. Corrupt and oppressive Churchmen produced a crisis within Christendom. Thomas Gascoinge, Chancellor of Oxford, noted in 1450 that "sinners say nowadays: 'I care not how many evils I do in God's sight, for I can easily get plenary remission of all guilt and penalty by an absolution and indulgence granted me by the Pope, whose written grant I have bought for four or six pence.'"[109]

The Reformation: The Defining Event of Modernity

Martin Luther (1483–1546), in 1517, affixed to the church door in Wittenberg his *Disputation on the Power and Efficacy of Indulgences*, later known as the *Ninety-Five Theses* that initiated the Reformation. Three principles defined the Protestant Reformation: The Bible as the sole authority, justification by faith alone, and the priesthood of all believers. All three served to substitute the individual for the community. Instead of submitting to the Church's explanations of the Bible, the Protestant turned to private interpretation. Instead of relying on the spiritual direction of priests and the prayers of Church members, each Protestant struggled alone for his or her salvation, not through any form of intercession but by individual effort and personal faith in Christ the Savior. All believers were equally thought to be priests. "For whoever comes out of the water of baptism can boast that he is already a consecrated priest, bishop, and pope," Martin Luther taught.[110] In effect, each Protestant became his or her own church.

"The quintessentially modern idea of the individual — and of one's personal responsibility before one's self and God rather than before any institution, whether church or state — was as unthinkable before Luther as is color in a world of black and white," writes Eric Metaxas in his biography of Luther.[111]

The Reformation launched individualism and thereby destroyed medieval communal life. Protestantism replaced the community by the individual; the new man of God was to achieve "salvation through unassisted faith and unmediated personal effort."[112]

In the New World, the Puritans put an indelible stamp of individualism upon America. After a livelong study of Puritanism, historian Perry Miller writes, "The Protestant sense of the individual, of the single entity which is one entire person, who must do everything of himself, who is not to be cosseted or carried through life, who in the final analysis has no other responsibility but his own welfare, this ruthless individualism was indelibly stamped upon the tradition of New England . . . . Both in economics and in salvation, the individual had to do everything of himself."[113]

Tocqueville captured in one word the essence of Modernity. He was the first person to use the word "individualism" and reports "that word 'individualism,' which we coined for our own requirements, was unknown to our ancestors, for the good reason that in their days every individual necessarily belonged to a group and no one could regard himself as an isolated unit."[114] The Latin word "*indīviduum*," the root of the English word "individual," *means an indivisible whole existing as a separate entity.*

Stated in its most general form, the defining principle of Modernity is that every whole — a political community, a horse, or a carbon atom — is a sum of its isolated parts. We will call this principle individualism, a straightforward extension of Tocqueville's original meaning and of Descartes' rule to begin with the parts. An excellent example of Cartesian reductionism is the two-fold individualism of Neo-Darwinism: First, an organism is considered an isolated individual separate from its parents and community; second, the organism is regarded as a collection of individual traits that are subject to natural selection.

Modernity rests upon three legs: science and technology, democracy, and capitalism; all three legs are bound together by individualism. The overarching principle of Modernity, then, is that *things exist in isolation* as separate entities.

In Old Europe, like in every premodern culture, the group was considered prior to the individual in origin and authority. Jacob Burckhardt, the great scholar of the Italian Renaissance, explains that in Medieval Europe, a "man was conscious of himself only as a member of a race, people, party, family, or corporation."[115] When asked "Who are you?", a person may have replied, "A Vignola from Padua, a stone carver, and a good Christian."

In Medieval France, the basic unit of society was the peasant family, the *domus*; the Latin word meant both family and house, for the two were inextricably bound together. The inhabitants of small villages seldom used the word *familia*; for peasants the "family of flesh and blood and the house of wood, stone, or daub were the one and the same thing," according to historian Emmanuel Le Roy Ladurie.[116] The central elements of the house were the kitchen fire, goods and lands, children, and conjugal alliances with other *domūs*. The *domus* usually went beyond two parents and their children to include servants, boarders, and illegitimate children, if any.

Sociologist Robert Nisbet concurs that in Medieval Europe, "the group was primary; it was the irreducible unit of the social system at large. The family, patriarchal and corporate in essence, was more than a set of interpersonal relations."[117] Taxes and fines were levied upon the medieval family, not the individual. Honors of achievement were bestowed upon the family, rather than the individual. Property belonged to the family, not the individual, and could not easily be separated from the family. The legal rights of the family over its members were inviolable. The family made almost all decisions affecting a person's occupation, marriage, and the rearing of his or her children.

Since the whole is seen as prior to and greater than any of its parts, the overarching principle in Medieval Europe as well as in all premodern cultures is *things exist only in relationship*. In Buddhism, a flower or a lion is said to be empty, meaning that the flower or lion has no independent, separable existence. Aristotle takes as obvious that "man is by nature an animal intended to live in a polis ... [as shown] by the faculty of speech."[118] Without a polis, a man is "either a beast or a god."[119]

The intellectual habit of modern Westerners seeing a part isolated from the whole has practical consequences. For example, chlorofluorocarbons, used as refrigerants and aerosol propellants, were tested extensively on human beings and found to be harmless. But the tests were carried out on individuals in laboratories, isolated from the environment. Years later, it was discovered that when chlorofluorocarbons are used by real people in the real world, the molecules migrate to the upper atmosphere and deplete the ozone layer. An ozone-depleted atmosphere allows more ultraviolet radiation to strike the surface of the earth and thus increases the incidence of cataracts and skin cancer. The original tests were failures because they did not take into account that human beings are part of an ecosystem. Because Westerners overlook the whole, they repeat the same basic error again and again.

The ancient and modern ways of understanding humans, nature, and the transcendent differ radically. In Modernity, the cosmos is "opaque, inert, mute"[120], unlike the ancient outlook, as developed by Aristotle and Aquinas, where *Homo sapiens*, an integral part of nature, shares a life with plants, animals, and the Prime Mover (God), all of which form a hierarchy ordered by degrees of nonmateriality. While self is not ignored by these two ancient thinkers, they emphasize the soul, what is universal about each person. In Modernity, of course, the soul is replaced by the isolated, autonomous self.

We have arrived at the Great Chasm that separates modern and ancient cultures, a chasm that may be bridgeable intellectually but not experientially. We moderns live in a totally different cosmos than our medieval ancestors or our Greek forbearers. Aristotle believed that the stars traverse circles about the Earth because of their desire to emulate the Prime Mover, an eternal being beyond the sphere of fixed stars that moves as an object of love, yet itself is unchanging.[121] Aristotle inhabited a tiny, comforting cosmos, strange to us, thanks to the truly glorious scientific revolution initiated by Copernicus. We cannot go home again to the cozy, ancient cosmos, where the night sky displayed the transcendent and Mother Earth manifested harmony and fecundity. Nor can we undo scientific knowledge; we live on a tiny planet, orbiting an ordinary star, near the edge of an ordinary galaxy that contains at least two hundred billion stars, in a universe with more than a hundred billion galaxies.

Aquinas believed the Garden of Eden existed in the East and that the location of Paradise was "shut off from the habitable world by mountains, or seas, or some torrid region, which cannot be crossed; and so people who have written about topography make no mention of it."[122] Unlike the theologians, saints, and peasants of Medieval Europe, we are not anchored to a narrow tradition ignorant of Buddhism, Hinduism, and Taoism. For us, living in a world of many differing cultures, the spiritual life must include the deepest insights of all wisdom traditions.

Unfortunately, we moderns are on the wrong side of the chasm, for the principle *things exist in isolation* is false. The ancient principle that *things exist only in relationship*, however, is very much present in everyday modern life, contrary to our defining cultural myth.

6 Things Exist Only in Relationship

The core tenet of Descartes' philosophy and thus of modern thinking is to begin "with the simplest and most easily known objects in order to ascend little by little, step by step, to knowledge of the most complex."[123] This sacred rule is clearly false since a part in biology cannot be understood in terms of itself. To have our thinking correspond to the way things are, we must replace the cultural dictate *to understand something, first understand its isolated parts* with the principle *to understand something, see how it is related to the whole.* Nothing exists in isolation.

Instead of beginning with the parts in my habitual manner, I forced myself to begin with the whole and refused to intone the Cartesian sacred hymn "begin with the parts," but instead chanted the new mantra "begin with the whole." Since reductionism was so thoroughly ingrained in my thinking, I needed numerous examples of the principle that things exist only in relationship.

Matter

In the nineteenth century, physicists hoped that someday they could isolate the atom from the cosmos, for they believed that knowing the properties of isolated atoms was the key to understanding the material world. Physicists later, however, discovered that the more an atom is isolated, the less actual it is. Atoms and elementary particles do not exist in the same way that billiard balls and cue sticks do. Atomic entities exist as potentialities or possibilities rather than as definite, concrete objects. In the twentieth century, quantum physicists were forced by nature to renounce the cultural dogma that the world is made up of autonomous parts, each with a separate, independent existence. Renegade physicist David Bohm sums up the essential feature of quantum physics: "The primary emphasis is now on *undivided wholeness*, in which the observing instrument is not separated from what is observed."[124] (For a fuller discussion, see Chapter 11: Quantum Physics and Mind.)

Nothing has an independent existence separable from everything else. *Things exist only in relationship* — and, this has always been true, even in Newtonian mechanics, although physicists for over two centuries unwittingly promoted the fiction that the world is made up of separate, independent parts.

In the daily work of science, physicists, mathematicians, and astronomers apply Newton's three laws to idealized objects that exist by themselves in an imaginary universe. Often professors and students alike take what is constructed for mathematical convenience as reality. Such idealizations often fail to capture the interconnectedness of nature. Physicist Richard Feynman, for example, demonstrated that "even simple and idealized things, like the ratchet and pawl, work . . . in only one direction because it has some ultimate contact with the rest of the universe."[125] He showed that if a mechanical watch were in a box, isolated from the universe, the heat buildup from friction would eventually cause the watch to keep time chaotically. For a watch, an automobile, or an electric motor to keep running in one direction, it must dump the heat it generates into its surroundings, and at some point, this requires that the heat generated on Earth be radiated into empty space. The Earth can cool off only because the universe is expanding and cooling down. A watch can keep time because the Big Bang started the universe in a one-way direction. For a physicist to understand completely why a machine can run in only one direction, she must understand the Big Bang.

Sense Experience

We must not be misled into thinking that *things exist only in relationship* applies exclusively to the exotic realms of quantum physics and cosmology. If we were raised from infancy as isolated individuals, we literally could not understand what we see. For human vision to be meaningful, a person must be a participant in the world. This surprising property of vision was demonstrated in a series of classic experiments by Theodor Erismann.[126] He fitted persons with vision-distorting goggles that made straight lines appear curved, right angles seem acute or obtuse, and distances seem expanded or shortened. Amazingly, after a few days, a subject's vision was no longer distorted; he saw normally and functioned normally, even skiing and riding a motorcycle!

The key to vision returning to normal was that the subjects were allowed to move about and act freely, enabling the strange new visual data to be integrated with the subject's experience of self-movement and self-sensation through touch. Subjects not allowed to move on their own, though they were pushed on gondolas through the environment, never experienced normal vision while wearing the distorting goggles.[127] To see the world, we must be participants, not mere spectators.

The senses are meant to be engaged with the outside world and the mind with something other than its own thoughts. In isolation, the senses and the mind create phantoms. Experiments on human subjects in isolation tanks demonstrated that extreme sensory deprivation induces such psychic disorders as mental confusion, hallucinations, and panic.

A human being exists only in relationship. Perceiving, feeling, imagining, thinking, and willing are impossible in isolation. A person in isolation from a larger whole, say nature or community, is a meaningless abstraction, an idealization that can only occur in philosophy and political theory. The isolated, autonomous self is a cultural myth whose realization would reduce a person to nothingness.

Family

Social psychologists Hazel Rose Markus and Shinobu Kitayama confirm that human beings exist only in relationship: "Persons are only parts that when separated from the larger social whole cannot be fully understood. Such a holistic view is in opposition to the Cartesian, dualistic tradition that characterizes Western thinking and in which the self is separated from the object and from the natural world."[128] In their convoluted, social science prose, Markus and Kitayama agree with my new mantra, "begin with the whole."

If we could sever all our ties to nature, family, and community, then we would cease to be. For example, the DNA that each of us bears in every cell of our bodies came from our parents, half from our mother and half from our father. If we tried to remove every trace of parents from our lives, we literally would not exist.

The self exists only when connected to others. Members of a family share the same hopes, the same joys, the same sorrows, and the same experiences; each family member lives a common life, each a part of the others. Divorce severs certain legal obligations, not the ties between spouses and their children, which are inseparable. For better or for worse, a common life yokes persons of the same family together forever. We do not live separate, parallel lives; we are not separate, isolated selves; each member of a family is a part of the others.

A former student of mine, Maggie Burnham, told me, "My parents were just two people who happened to live in the same house as me. Whenever they forbid me to do something, I would be furious that they had the nerve to do such a thing." When her mother suddenly became ill with ovarian cancer and died, the daughter felt a great loss in her life.

7 Self-Interest: The Other Pathology of Modernity

Every empire is run by a professional class schooled to administer the daily affairs of the state. For instance, Hammurabi established the Babylonian Empire in 1772 B.C., and soon, special schools were created to rigorously train scribes, librarians, and accountants, the world's first professional class, whose duty was to record everything consumed in the temples, to keep lists of kings and dynasties, and to chronicle significant historical events.

John le Carré, the acclaimed author of *The Spy Who Came in from the Cold*, explains why the education of upper-middle-class Brits was perfect for governing the British Empire: "For our class in my era, public school was a deliberately brutalizing process that separated you from your parents, and your parents were parties to that. They integrated you with imperial ambitions and then let you loose into the world with a sense of elitism — but with your heart frozen." His friend Ben Macintyre, also a writer of espionage novels, adds, "There is no deceiver more effective than a public-school-educated Brit. He could be standing next to you in the bus queue, having a Force 12 nervous breakdown, and you'd never be any the wiser."[129]

In the era of le Carré and Macintyre, the graduates of public schools such as Eton College, Charterhouse, and Harrow School were charming, highly competitive, emotionally distant, indifferent to the pain of others, and deceitful possessors of frozen hearts perfect for the professionals needed to rule Britannia, to rule the waves, and to take up the White Man's burden.

Many Americans refuse to acknowledge that in the twentieth century after the British, French, and Dutch were forced to abandon India, Africa, and Southeast Asia, the United States became by default an empire; however, virtually, no one doubts that America's contribution to humankind is material prosperity for all founded on political freedom, technological innovation, and free markets, in effect, an empire of consumer goods and physical comfort.

We Are Schooled for Capitalism

Public, private, and parochial schools are founded on grading, a system of competition that instills the ethos of capitalism. The most lasting lessons in competition are given in the classroom, not in the workplace or on the playing field.

Anthropologist Jules Henry describes how competition entered into a fifth-grade arithmetic lesson he observed: "Boris had trouble reducing 12/16 to the lowest terms, and could only get as far as 6/8. The teacher asked him quietly if that was as far as he could reduce it. She suggested he 'think.'" Undoubtedly, Boris remembered hearing the teacher tell him to reduce the fraction to the lowest terms, but then he could not speak. When the teacher told him to think, his mind was probably paralyzed, and his ears buzzed. Other children, frantic to correct Boris, waved their hands to get the teacher's attention. The teacher, quiet and patient, ignored the waving hands and asked Boris, "Is there a bigger number than two you can divide into the two parts of the fraction?" After a long silence from Boris, she asked the same question again, this time more urgently, and still there was not a word from Boris. She then turned to the class and said, "Who can tell Boris what the number is?" A forest of hands appeared, and the teacher called on Peggy, thrilled to give the correct answer, four. Both Peggy and Boris see the classroom as a place where they are told to do certain things and praised if they do them right, disapproved if they do not.[130]

Henry points out that the grade-school lesson in competition he witnessed is the "standard condition of the American elementary school." From the smile on her face, Henry knew Peggy felt great about herself; Boris' failure was his problem, not hers. In the schoolhouse, winners are taught to look to the good they have gained and ignore the unavoidable, emotional damage caused to the losers. The goal in a competitive society is to win without violating the rules. That is how the game works in America, and that is how the natural empathy young children feel for the pain of others is squashed by the ethos of capitalism. Later in life, Peggy and most of her classmates out of "ignorant and coarse"[131] self-interest learned in grade school would be indifferent to the fate of the three million children in America who live in abject poverty, the kind found in Bangladesh, one of the poorest countries in the world.[132] Education induces blindness to the 53 million low-wage workers earning median hourly wages of $10.22[133] and to the workers riding the down escalator to the underclass.

Contrary to their social nature, the unsuccessful students grow to hate the successful ones. "Since all but the brightest children have the constant experience that others succeed at their expense, they cannot but develop an inherent tendency to hate — to hate the success of others, to hate others who are successful, and to be determined to prevent it. Along with this, naturally, goes the hope that others will fail. This hatred masquerades under the euphemistic name of 'envy.'"[134]

Ten years later, probably few of the students that Henry observed will remember how to reduce fractions. But surely no student will forget the real lessons learned that day, the three moral precepts of capitalism. *I succeed only if someone else fails*, and the converse — *if someone else succeeds, I must have failed.*

Implicit in these two precepts is a third: *My success is entirely due to me, and no other person has a legitimate claim on its benefits* — the ultimate injunction of capitalism, an economic system based on individualism, where each person is solely responsible for his or her success or failure.

Education critic John Holt contends that competition in school destroys the "love of learning in children, which is so strong when they are small, by encouraging and compelling them to work for petty and contemptible rewards — gold stars, or papers marked 100 and tacked to the wall, A's on report cards, or honor rolls, or dean's lists, or Phi Beta Kappa keys — in short, for the ignoble satisfaction of feeling that they are better than someone else."[135]

Grade-school students do not have an inkling that they are being prepared for the workplace, where "the isolated individual has to fight with other individuals of the same group, has to surpass them and, frequently, thrust them aside," observes psychoanalyst Karen Horney. "The advantage of the one is frequently the disadvantage of the other." The situation where everyone is a real or potential competitor of everyone else creates a diffuse hostile tension between individuals, as is clearly apparent among members of the same occupational group, regardless of the disguised attempts to camouflage envy and hatred by politeness. "Competitiveness, and the potential hostility that accompanies it, pervades all human relationships," Horney concludes from her years of psychoanalytic practice.[136] Psychotherapist Rollo May agrees: "Individual competitive success is . . . the dominant goal in our culture."[137]

Despite the introduction of cooperative learning, the awarding of trophies to every member of a Little League team, and the inflation of grades, not that much has changed in American education since its critique by Henry and Holt fifty years ago. Arguably the No Child Left Behind program of the Bush administration and President Obama's Race to the Top, with their almost exclusive emphasis on testing and meeting national standards, have increased competitiveness among students and between individual teachers.

Instead of fostering their social nature, grade school teaches students to compete with their classmates. School-wide activities with a common purpose are absent from most schools. A common task such as cleaning the schoolyard would give students the experience of working for a common end, and when the end was achieved, each of them would feel joy. In this way, they would learn that a person's good and group's good can coincide.

To succeed in a combative society, students are told that they must be competitive and rivet their attention on self-interest. Grade-school character training instills "social virtues," such as selfishness, aggressiveness, and competitiveness. From grade school to graduate school, it is pounded into students' heads that those without these strengths will lose in the struggle of life, will sink into a second- or third-tier life, and that they need not worry about anyone but themselves. If each person looks out for his or her own welfare, in the end, everyone will make out all right.

Given their schooling for capitalism and their enthrallment with the celebrityhood promoted by mass media, graduates from high school and college believe wealth and fame are the two keys to the good life. In a recent survey of young adults, over 80 percent said that a major life goal was to get rich, and 50 percent of those same interviewees stated that another major life goal was to become famous.[138] To obtain happiness for themselves, many young adults intend to fight their way to the top and let the rest of the world be damned, a course of action contrary to the finding of the Harvard Study of Adult Development, a research project that since 1938 closely tracked the lives of 724 men.

Robert Waldinger, the fourth director of the study, summarized in a Ted Talk the result of over 75 years of empirical investigation of adult development: "Good relationships keep us happier and healthier. Period."[139] He gave two other major lessons learned from the study. Loneliness is toxic, causing physical and mental declines in midlife; loneliness and weak social connections are associated with a reduction in lifespan. "The sad fact is that at any given time, more the one in five Americans will report that they are lonely," Waldinger lamented.

The other surprising lesson from the study is that at midlife, cholesterol levels were not as a good predictor of a healthy old age as the quality of human relationships. "Good close relationships seem to buffer [a person] from the slings and arrows of growing old." People in close relationships where they can count on the other person in times of need are happier, healthier, and experience less memory decline than those in high-conflict marriages. The study concluded from 75 years of empirical data that the most important predictors of living a long and happy life are not money and notoriety but the strength of relationships with spouses, family, and friends — in a word — love.[140]

In today's mail, I received from a former student a large manila envelope. On the back of which he had written, "Inside is a postcard that my son's third-grade teacher taped to his classroom door for a week. I hope you enjoy this new Credo for Young America." With the penknife that is always on my desk, I slit open the envelope and pulled out a 4x6 postcard. (See figure 7.1) I slowly read several lines of "poetry" printed on the card.[141]

I could hardly keep from laughing. In the "poem," the grade-school language from my days in elementary school now included the buzz words of psychotherapy; yet, the message remained the same — "I am my own private property." Pre-school children learn the phrases "my toys," "my clothes," "my money," and "my room" before mastering the alphabet. To further their education in capitalism, elementary school children are taught to understand themselves as a private possession that no one has any claims on.

In America, students are taught that they are unique and that they alone choose their feelings, dreams, and hopes — everything that defines them. In reality though, they are much like premodern Africans, whose opinions and judgments were formed by their mothers, elders, and teachers.[142] Ethnologist

S. M. Molema reports that in his tribe, the Bantu, a person's *"very words are often a formula,"*[143] not unlike Virginia Satir's words to American grade school students; few as adults will wake up to see that they have been speaking formulas for years.

MY DECLARATION OF SELF-ESTEEM

I AM ME

In all the world, there is no one else exactly like me.

Everything that comes out of me is authentically mine because I alone choose it.

I own everything about me: my body, my feelings, my mouth, my voice, all my actions, whether they be to others or to myself.

I own my fantasies, my dreams, my hopes, my fears.

I own all my triumphs and successes, all my failures and mistakes.

Virginia Satir

Figure 7.1 Virginia Satir's words to American grade school students.

8 How Technology Begot Materialism

Before we examine if scientific evidence supports the opinion that the human person is a material being, let us turn to how materialism came to dominate popular thinking.

Nonscientists did not follow the path of reductionism to materialism. Most citizens of Orange County, California, or any other place in this country, do not think of themselves as a bundle of simple quarks (what the hell are those?) and electrons (those run through the electrical wires in my house, right?), unless they are insane. The great mass of inventions founded on modern science drew ordinary people to materialism.

Jacques Vaucanson, in 1738, exhibited in Paris a mechanical duck that ate, digested food, and defecated. The clockwork mechanism of the duck consisted of numerous levers and interlocking gears; each wing alone had more than four hundred moving parts. The inventor described his duck as an animal "made of gilded copper, who drinks, eats, quacks, splashes about in water, and digests food like a living duck."[144] The duck took kernels of corn from Vaucanson's hand and swallowed them with a quick, realistic gulping action of its flexible neck. Inside the duck, a "chemical laboratory" digested the food. The chyme passed into the bowels, then to the anus, where a sphincter permitted it to emerge. (See Figure 8.1.)

Vaucanson's duck was a sensation in Paris. To see the mechanical marvel of the age, poor people paid a week's wages. The great Voltaire pronounced the digesting duck and its inventor a fulfillment of humankind's ambition to create life and thereby displace the gods: "A rival to Prometheus, Vaucanson seemed to steal the heavenly fire in his search to give life."

Forty-five years later, the magician Eugéne Robert-Houdin (the conjuror from whom Houdini took his name) discovered the duck was a fraud. The corn eaten by the duck, instead of being digested, was simply stored in a container at the base of the throat. The substance emitted from the anus was pre-stored "feces," dyed green breadcrumbs, pumped out of the duck and in the public demonstrations by Vaucanson collected with great care on a silver platter.

Despite the exposure of Vaucanson's duck as a conjuring trick designed to convince the rubes that life is a mechanical phenomenon, the Cartesian image of nature as a clockwork mechanism seized the imagination of Western Europe.

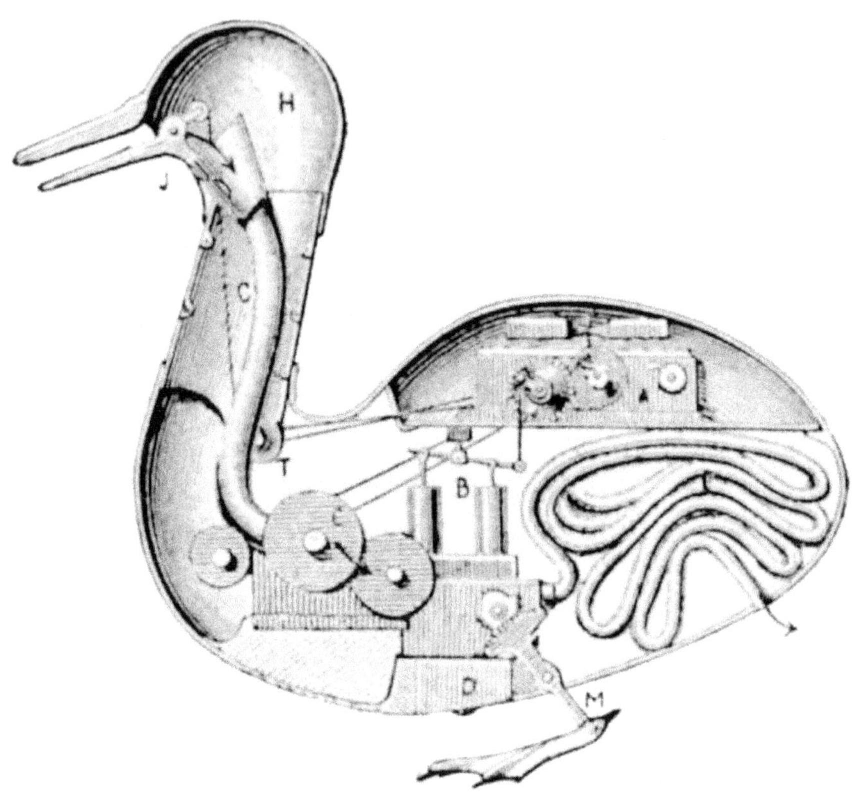

Figure 8.1. A mistaken representation by an American observer of how the Defecating Duck worked. (Public domain, copyright expired, Wikimedia Commons.)

In Modernity, life is believed to be a mechanical phenomenon, much like Vaucanson's duck. Rain falls; water seeps into the ground; its absorption by a seed starts a chemical reaction; the seed sprouts, eventually sunlight impinges upon the new leaves; the resulting photochemical reaction supplies the energy for further growth. In a similar mechanical manner, the plant reaches maturity, reproduces, and dies. An animal's development fits into the same mechanical scheme; a fertilized ovum replaces the seed and food plays the part of sunlight.

Before Modernity, nature was understood in an entirely different manner. For example, in medieval theology, plants, animals, and human beings are linked by the desire to return to God, a Christian version of the natural philosophy of Aristotle. As we saw in Chapter 1, Aristotle concluded from his study of nature that plants and animals possess an inner agency that causes them to emulate the Prime Mover; every plant and animal desires full actuality, which causes seeds and fertilized ova to grow to adult organisms. Furthermore, each plant and animal desires immortality, but the corruptibility of matter forbids that, so a plant or an animal settles for what is second best, reproduction of itself to continue its species, and that explains the cycles of generation and corruption found throughout nature. The great principle of nature is not mechanical causation but love. Every living organism yearns for full existence, forever.

Fortuna, an old Indian woman from the San Ildefonso Pueblo, repeatedly told me that nature reveals the supernatural; a rock is never merely a rock, or a bird just a flying machine, or a human being an animal that appears between one nothingness and another. Nature always expresses the transcendent. But for moderns, nature is opaque, mute, transmits no message, and holds no key to existence, and this is true even for the Aristotelian philosophers I know, for whom nature appears in ancient treatises not experience.

The "great mass of inventions"[145] predicted by Bacon to flow from science begot the Machine Age. A quick check of the "Timeline of Historic Inventions" in Wikipedia shows a rapid acceleration of inventions after Bacon argued in his *The New Organon* (1620) for founding science on experiment. Before 1620 the number of inventions is meager. The entry for the sixteen century has eleven inventions, including scissors, ether, the musket, and the pencil; the entry for the century in which Newton died, the eighteenth, has thirty-five entries, including the tuning fork, the flying shuttle, the Franklin stove, the Watt steam engine, the steamboat, bifocals, the cotton gin, and vaccination. In the nineteenth and twentieth centuries, inventions are so numerous they are listed by decades. The 1920s saw the polygraph, television, sound film, aerosol spray, sliced bread, and the extraction of insulin from animals for the treatment of diabetes.

From the eighteenth century on, no person in the Western World could fail to see that scientists commanded nature. Even the most poorly educated peasant could understand that the state of scientific knowledge was greatly advancing, while the most sophisticated intellectual knew that philosophy remained a domain of "contentious and barking disputation,"[146] where nothing is ever settled. Smallpox vaccine, the steam train, and the electrical telegraph demonstrated that science was the true path to knowledge, and that was bad news for theologians, philosophers, and poets, for they could not command nature. Their prestige began a steady, irreversible decline. Today, no one speaks of poetic knowledge, philosophical insights into nature and human affairs, or killer arguments for the existence of God.

The amazing inventions flowing from science focused everyone's attention on the good life in this world, at the expense of intellectual and spiritual values, the usual definition of materialism.

9 Dumb Idea # 3: Materialism

Trained to believe that every object as well as every act in the universe is matter, an aspect of matter, or produced by matter — that is, schooled to be a materialist — I scoffed at the two fellow students of mine in graduate school who regularly attended church. For me, at that time, the brain was the mind and God an illusion.

Sunday Morning in the Cathedral of Science

Figure 9.1. Wilson Hall, the "High Rise," at Fermilab, Batavia, Illinois. Courtesy Fermi National Accelerator Laboratory

Seated in the front pew, my folded hands piously resting upon a worn copy of Newton's *Principia*, I hear from the choir loft the voices of neuroscience graduate students droning their mantra, "The brain is the mind; The brain is the mind; The brain is the mind."[147] The mantra becomes the astonishing hypothesis in the Sunday morning sermon preached by the Reverend Francis Harry Compton Crick, the co-discoverer of the double-helix structure of DNA: "'You: your joys and your sorrows, your memories and your ambitions, your sense of personal identity and free will, are in fact no more than the behavior of a vast assembly of nerve cells and their associated molecules." With index finger pointing heavenward, Reverend Crick bellows the crux of the sermon: "You're nothing but a pack of neurons."[148]

In a monophonic chant that hypnotizes the parishioners, psychologist Joshua Greene recites the liturgical reading of the day: "Every decision is a thoroughly mechanical process, the outcome of which is completely determined by the results of prior mechanical processes."[149] Every human action can be explained mechanically. In a higher octave, biologist Lynn Margulis trumpets, "For all our imagination, fecundity, and power, we are no more than communities of bacteria, modular manifestations of the nucleated cell."[150] Evolutionary biologist Richard Dawkins recites the second reading, "[Replicators] swarm in huge colonies, safe inside gigantic lumbering robots, sealed off from the outside world, communicating with it by tortuous indirect routes, manipulating it by remote control. They are in you and in me; they created us, body and mind; and their preservation is the ultimate rationale for our existence. They have come a long way, those replicators. Now they go by the name of genes, and we are their survival machines."[151]

Physicist Robert Cahn reads from the Book of the Standard Model, "Given the masses of quarks and leptons, and nine other closely related quantities, [the current theory of particle interaction] can account, in principle, for all the phenomena in our daily lives."[152] In the same vein, Murray Gell-Mann, the first theoretical physicist to propose that the fundamental building blocks of matter are not protons, neutrons, and electrons, but quarks and leptons, intones, "All of us human beings and all the objects with which we deal are essentially bundles of simple quarks and electrons."[153] In a dark corner of the Cathedral of Science, in *sotto voce* Stephen Hawking murmurs, "The human race is just a chemical scum on a moderate-sized planet."[154] (See Figure 9.2.)

The uninvited preacher, Marvin Minsky, a leading proponent of artificial intelligence, shouts from the last pew in the Cathedral of Science, "Who are we? We are 'meat machines'."[155]

Not one grand pronouncement preached that Sunday in the Cathedral of Science has been established by experimental science. "You're nothing but a pack of neurons" and "human beings . . . are essentially bundles of simple quarks and electrons" are for the majority of scientists unshakeable beliefs, ultimately religious dogmas.

Figure 9.2. What caused this chemical scum to wash up on shore? Jan-Erik Finnberg, *Coney Island Beach Full of People*, detail, Flickr Creative Commons.

Every sermon, every reading, and every mantra in the Cathedral of Science rests upon the Central Dogma: The universe, including all aspects of human life, is the result of the interactions of little bits of matter.

The New Reality of Science: The Nonmaterial

The program to prove the *Homo sapiens* is a material being is an absurd, outmoded agenda of the Dark Ages of late modern science. The hope of neuroscientists to show the brain is the mind by explaining how the interior life of the human being — perceiving, feeling, imagining, willing, and thinking — results from action potentials, the endocrinology of neurotransmitters, and the interactions of neurons is an impossible, mad dream.

The human brain contains 86 billion neurons, or nerve cells. A large number of little branches, known as "dendrites," extend from the cell body of the neuron and receive signals from other neurons, while an axon conducts neural messages to other neurons. (See Figure 9.3.) Each neuron makes electrical connections, or synapses, with as many as 10,000 other neurons. The brain of a three-year-old child has approximately one quadrillion synapses. A quadrillion is roughly the number of people on 140,000 Earths.

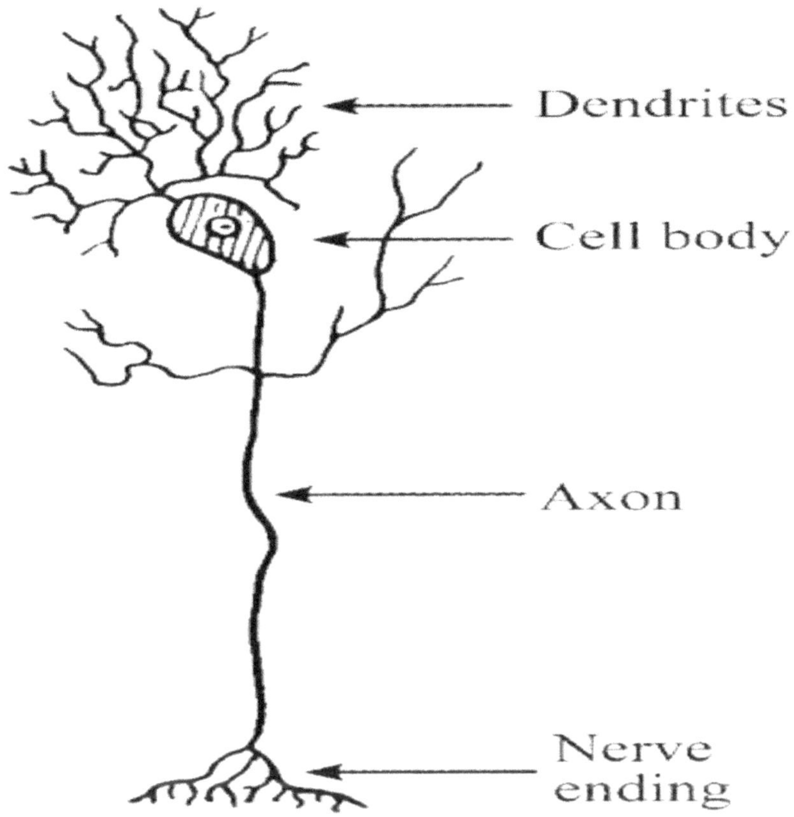

Figure 9.3. A schematic representation of the neuron with its main parts: the cell body; the main output fiber (the axon); the input fibers (the dendrites). Every neuron is different, just as in a forest no two pine trees are the same. (Pearson Scott Foresman, Public Domain, Wikimedia Commons.)

One seemingly insurmountable obstacle for a neuroscientist intent upon reducing a person to "nothing but a pack of neurons" is to explain how wiring together 86 billion neurons can give rise to the love of Mozart's *Don Giovanni* or to the joy of windsurfing at Maalaea Bay, Hawaii.

Brain Function Alone Cannot Explain Perception

I was astonished to discover that a careful analysis of perception, the bare minimum of human living, quickly shows that materialism is dead wrong. The first step in applying the dogma *the brain is the mind* falters; brain physiology alone cannot explain the most obvious human experience — we perceive. Textbooks typically gloss over the profound difference between sense perception and its necessary physical components. With regard to vision, Crick confesses, "We really have no clear idea how we see anything. This fact is usually concealed from the students who take such courses [as the psychology, physiology, and cell biology of vision]."[156]

Physicist Erwin Schrödinger argues that science is incapable of explaining how we see: "If you ask a physicist what is his idea of yellow light, he will tell you that it is transversal electromagnetic waves of wavelength in the neighborhood of 590 millimicrons. If you ask him: But where does yellow come in? He will say: In my picture not at all."[157] Carl von Weizsäcker, also a physicist, agrees: "Light of 6,000 Å wavelength reaches my eye. From the retina, a chemicoelectrical stimulus passes through the optical nerve into the brain where it sets off another stimulus of certain motor nerves, and out of my mouth come the words: *The apple is red*. Nowhere in this description of the process, complete though it is, has any mention been made that I have had the color perception *red*. Of sense perception, nothing was said."[158]

Although Schrödinger and Weizsäcker employ technical language, their insights into the nature of perception are based on straightforward observations. Let me restate their argument. Suppose sunlight is reflected from a red apple into the eye of a landscape painter. The sunlight passes through the lens of the eye and strikes the retina, a sheet of closely packed receptors — 4.5 million cones and 90 million rods. Activated by the incoming sunlight, chemical changes occur in the rods and cones, which are then translated into electrical impulses that travel along the optic nerve to the brain. Further electrical and chemical changes take place in the brain. In terms of the physiology of seeing this description is complete; however, the sensation *red* has not entered into this scientific account of perception. The landscape painter experiences the *red* of the apple, not the myriad chemical and electrical changes that are necessary for seeing.

What is true for seeing is true of the other senses: Physical-chemical changes in the brain are insufficient to explain any sensory perception. The sensible qualities a person perceives never appear in the brain as such. The brain itself is shrouded in complete silence, even while a person hears the deafening roar of a jet aircraft engine. Likewise, the brain, encased in the skull, is covered with darkness even while a person perceives the brilliance of the sun's glare. Our brains do not become colder when we touch snow or harder when we touch iron. The brain tissue itself takes on none of the sourness of a tasted lemon or the acrid odor of the skunk's spray that we smell.

Charles Sherrington, the founder of modern neuroscience, points out that physics and chemistry "bring us to the threshold of the act of perceiving, and there to bid us 'goodbye.'"[159] In the scientific picture of the world, colors, odors, sounds, flavors, and textures are absent. Thus, Schrödinger concludes that scientific "theories are easily thought to account for sensual qualities; which, of course, they never do."[160]

Mechanical, chemical, and electrical changes themselves are not thoughts, desires, and emotions. The toolbox of physical science is limited to air pressure, chemical changes, electrical impulses in nerves, brain cell activity, and other measurable properties of matter. Science is mute about hearing, seeing, smelling, tasting, and touching. The reason is obvious: the interior life of a person is nonmaterial — perceptions, emotions, and thoughts cannot be touched, smelled, tasted, heard, or seen.

Emergentism

My physicist colleagues and all the neuroscientists I know, except for one, attempt to evade Schrödinger's argument by invoking the doctrine of emergentism. An emergentist recognizes the genuine novelty in nature at each successive level of organization and affirms the hierarchy of natural beings. Biologist Peter Medawar puts it simply, "Each higher-level subject contains ideas and conceptions peculiar to itself. These are the 'emergent' properties."[161] Life, sensation, and mind are considered to be among these. In his paper, "More Is Different," condensed-matter physicist Philip Anderson maintains that "psychology is not applied biology, nor is biology applied chemistry."[162] According to the doctrine of emergentism, when 86 billion neurons are wired together in the right way, consciousness emerges from the non-conscious basic elements; more is different.

Indeed, in physics, sometimes more is different. In Newtonian physics, for example, a particle moving through empty space experiences no drag. If a huge number of these particles are put together to form a fluid, and if these particles interact with each other, then every particle moving in the fluid experiences a drag that causes a stickiness of the fluid, called "viscosity." Clearly, viscosity is an example of where more is different. However, the motion of a single particle through empty space and the motion of a viscous fluid are commensurable, meaning they are both measurable in terms of mass, force, momentum, and acceleration, which is not true in the case of the "emergence" of perception, for as Schrödinger, Weizsäcker, and Sherrington point out air pressure, chemical changes, electrical impulses in nerves, brain cell activity, and other measurable properties of matter are not commensurate with smell of lavender, the sound of middle C bowed on a violin or the golden light that envelops the high desert of Northern New Mexico at sunset.

My physicist friends always point to superconductivity as the paradigm of emergentism. While superconductors were unexpected states of matter, eventually these surprising states of matter were understood through the quantum mechanical behavior of electrons, not through some new laws of matter. Ordinary resistive conductors and superconductors are commensurable; both are understood through the quantum mechanical behavior of electrons moving through a metallic lattice. At a low enough temperature, electrons form Cooper pairs, composite bosons that can condense into the same ground state, a superconducting state of the metal.

Emergentists believe that something analogous happens with neurons and perception. Suppose that when 86 billion neurons are assembled in the right way that the perception of Middle C or the smell of lavender emerges, then to rationally understand why this happens, there must be an underlying connection, or link, between neurons, neurotransmitters, and action potentials and the sound of Middle C and the smell of lavender, but no link exists, as one does with superconductivity. The whole point of the Schrödinger argument is that these realms are totally different with no causal connection.

Emergentists adhere to the unshakeable belief that "the universe, including our own existence, can be explained by the interactions of little bits of matter"[163] and as a result encounter an insurmountable barrier: The properties of matter are incommensurate with the experiences of the interior life. For an emergentist to overcome this barrier by simply declaring that from an assembly of 86 billion neurons consciousness emerges is magical thinking.

A Tour of Beethoven's Brain

To observe what is going on in Beethoven's brain, a neuroscientist shrinks himself down to the size of a nerve cell and purchases a ticket from Great Western Microtours. He sees the intricate chemical changes at the synapses, making possible the transmission of a nerve impulse from one neuron to another. As the microtour bus zips through the brain, he witnesses electron transport, ion interactions, and the furious chemical activity of enzymes within each cell. Nowhere does he see or taste the Rhine wine being enjoyed right now by the brain's owner, even though the tour guide points out a particular series of electrical impulses that correlate with those sensations. Nowhere does he see Beethoven's remembrance of his first meeting with Mozart. No emotions are discernible either. He never recognizes this particular spatiotemporal pattern of neuronal activity as anger at having gone deaf or that complex electrical pattern of nerve cells as joy or fear or despair. At every level, the neuroscientist meets only with the physical correlates of sensations and emotions, never with the experiences themselves; *correlation is not causation*. He cannot explain why Beethoven experiences vivid colors, pungent odors, or vibrant tones. This is because the interior life of a human person is nonmaterial and cannot be found in the brain.[164]

The Brain from the Outside

Despite the marvelous advances in noninvasive diagnostic tools in medical science, fMRI, PET, and SPECT scans measure only brain activity, not the mental activity of the subject. For neuroscientist and layman alike, the interior life of another person is unobservable. A colored photograph of a brain state created by a brain scanner (see Figure 9.4) is not the perception of an apple, any more than a photograph of an apple is the texture and taste of an apple.

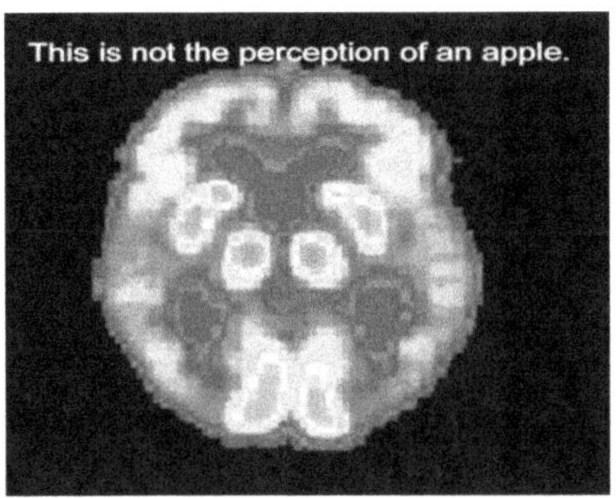

Figure 9.4. A brain scan of a subject viewing an apple.

The brain scan represents the perception of an apple in the same way that René Magritte wittily illustrated the truth that a representation is not the thing. (See Figure 9.5: *The Treachery of Images*, a realistic painting of a pipe on which Magritte inscribed "*Ceci n'est pas une pipe.*") However, some neuroscientists, victims of the treachery of images, think the scans of a brain record mental activity, not correlates.

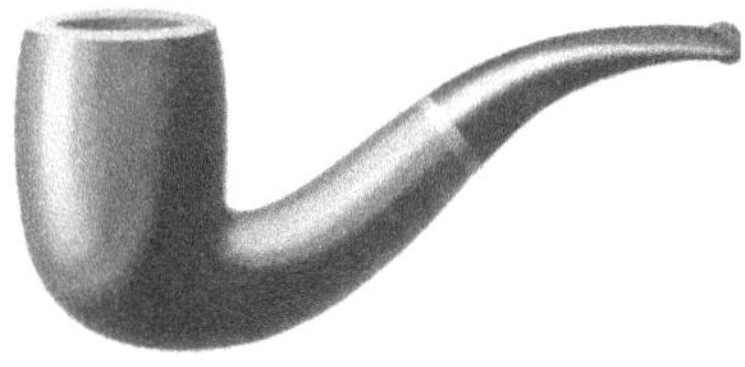

Figure 9.5 René Magritte, *The Treachery of Images*, 1928. Los Angeles County Museum of Art. Courtesy of Wikimedia Commons

Most scientists fail to see that the nonmaterial does not mean ethereal, spooky, or that the nonmaterial necessarily can exist separate from matter. The visual memories encoded in a person's brain are nonmaterial, but they do not exist separate from his brain. Neurologist Oliver Sacks reported that a patient under his care had suffered a sudden thrombosis in the posterior circulation of the brain, which caused the immediate death of the visual parts of the brain. The patient became completely blind — and did not know it! Sacks' questioning revealed that the patient had lost all visual images and memories, yet had no sense of loss. The patient had no memory of ever having seen; he was unable to describe anything visual and became bewildered when hearing the words "seeing" and "light." An entire lifetime of visual experience had been erased from memory in an instant.[165]

Encodements

Everyone knows that memories of seeing, hearing, tasting, smelling, and touching are stored in some manner in the brain, but what memories are encoded there cannot be read by scientific instruments; the neuroscientist must rely upon what the owner of the brain reports.

To understand why brain activity alone cannot account for a person's experience of remembering, we must look at what all encodements have in common. Consider Lincoln's "Gettysburg Address." It can be spoken in English or German, written in Arabic or Chinese, tapped out in Morse code, given in raised dots of paper (Braille), or encoded on audiotape by magnetizing the tiny iron oxide pieces embedded in the tape. Clearly, a code is arbitrary, not natural. There is no logical or necessary physical connection between a code and what is encoded. The meaning of "four score and seven years ago" is not sounds in English, black marks on paper, raised dots, or magnetic patterns. Without a decoder, the raised dots shown in figure 9.6 are meaningless — they are the first six words of the "Gettysburg Address" in Braille.

Figure 9.6 "Four score and seven years ago" in Braille.

"Four score and seven years ago" has two elements — the meaning and the medium of the encodement. The medium is material, while the meaning is nonmaterial. The meaning of the "Gettysburg Address" cannot be reduced to the mere arrangement of sounds in English or impressions on paper, nor can it be measured in terms of mass, length, time, and electric charge.

The memories encoded in the brain are analogous to the sounds recorded on an audiotape. Just as the magnetic patterns on an audiotape are *not* the "Gettysburg Address," the chemical and electrical patterns in the brain are *not* the memory of New York Governor George E. Pataki reciting the "Gettysburg Address" in a tribute to the victims of the 9/11 attacks on America. To go from the magnetic patterns on the tape to sound requires a decoder (a tape player). Similarly, no matter how sophisticated the study of a person's brain, neuroscientists can never read the memories stored there; the decoder, in this case, is a mysterious device embedded in the owner's brain.

In conclusion, we must point out that because materialism is a metaphysical belief — a cultural myth, a sort of intellectual creed — added on to science, its death does not alter any scientifically established result: Atoms still exist; the Earth is still 4.5 billion years old; and the universe still began with a Big Bang.

10 The Primacy of Mind

The Blind Hope in Materialism Persists

The hope that someday physics, chemistry, biology, and computer science will eventually explain mind lingers. Neurologist Antonio Damasio, for instance, believes the "simple sensory qualities to be found in the blueness of the sky or the tone of sound produced by a cello . . . will be eventually explained neurobiologically, although at the moment the neurobiological account is incomplete and there is an explanatory gap."[166] Neuroscientist Christof Koch maintains that all facets of consciousness — visual, olfactory, linguistic, even the self — are "elaborations of a common biological process," and in his laboratory, he is searching for "a discrete set of neurons that might be in twenty different areas [of the brain] but share some set of properties that are responsible for generating consciousness."[167]

At a public lecture sponsored by the Santa Fe Institute, I heard Francis Crick publicly admit that the inability of neuroscientists to explain sensory qualities in terms of brain activity seems insurmountable. Yet, he hoped that the problem would unexpectedly disappear, like other seemingly impossible difficulties had in science. In a joint paper, Crick and Koch ask, "What Are the Neural Correlates of Consciousness?" and answer, "So far no one has put forward any concrete hypothesis that sounds even remotely plausible."[168]

Neuroscientist Vilayanur Ramachandran, too, acknowledges that explaining perception in terms of brain function alone is an impasse for neuroscience and for materialism in general: "No matter how detailed and accurate [the] outside-objective description of color cognition might be, it has a gaping hole at its center because it leaves out"[169] the experience of redness and all other perceptions. He laments that the impasse results from a limit of present-day science: "Perhaps, science will eventually stumble on some unexpected method or framework for dealing with *qualia* — the immediate experiential perception of sensation, such as the redness of red or the pungency of curry — empirically and rationally, but such advances could easily be as remote from our present-day grasp as molecular genetics was to those living in the Middle Ages."[170]

Biologist H. Allen Orr agrees that "how such *mere* objects [as a brain and its neurons] can give rise to the eerily different phenomenon of subjective experience seems utterly incomprehensible." Despite this, he holds that materialism need not be rejected for two reasons. The first reason is "more a sociological observation than an actual argument. Science has, since the seventeenth century, proved remarkably adept at incorporating initially alien ideas (like electromagnetic fields) into its thinking." This opinion is essentially Crick's hope that the problem of *qualia* will somehow disappear in the future.

Orr's second reason for not conceding that materialism is a failed philosophy is based on a conjecture about why *qualia* and consciousness pose an insurmountable obstacle to present-day science: Our inability to see how certain material structures produce subjective experience is that our evolved, finite brains cannot "fathom the answer to every question that we can ask." Orr is willing to believe matter gives rise to mind, but scientists will never know how, a belief that violates the central tenet of modern science — a theory in some way must be open to experimental refutation. In this way, Orr exposed that his unshakeable belief that "the universe, including our own existence, can be explained by the interactions of little bits of matter"[171] is an irrational ideology.

What Damasio, Koch, Crick, and Orr cannot face and what Ramachandran suspects is that materialism is dead and cannot be resurrected. Undeterred by the advent of quantum physics and chaos theory in the twentieth century, neuroscientists persist in reading nature as a machine, a view inherited from Descartes and Newton, and thus cannot see that *qualia* along with all aspects of the interior life are the fundamentals of human life, not the chemical and electrical activity of the brain.

Despite theoretical claims to the contrary, neuroscience does not account for a person's interior life in terms of matter alone. No researcher begins with neurons, neurotransmitters, and action potentials and then explains how sensations and emotions arise from these basic elements. Furthermore, the interior life of a person is not open to observation. Instead, neuroscientists in practice take perceiving, feeling, thinking, and willing as *primary* and then attempt to *correlate* brain activity with what the human subject reports. The chemical and electrical activity in the brain is understood in terms of the interior life of the human being. Through its practice, neuroscience affirms that the mind is primary, not matter. The mantra chanted daily by many neuroscientists and certain philosophers — the brain is the mind — invokes a hopeless cause, conjures up magical thinking.

Neuroscientist Roger Sperry, the discoverer of the different roles of the left and right hemispheres of the brain, describes how when science is freed from a materialistic outlook, the mind is primary: "Mind and consciousness are put in the driver's seat, as it were: They give the orders, and they push and haul around the physiology and the physical and chemical processes as much as or more than the latter processes direct them. This scheme is one that puts mind back over matter, in a sense, not under or outside or beside it. ... The causal potency of an idea, or an ideal, becomes just as real as that of a molecule, a cell, or a nerve impulse."[172] With this outlook, brain physiology, neurons, and neurotransmitters are the physical structures that allow a painter to see nature, a novelist to explore the interior life, and, yes, a neuroscientist to understand the brain.

When neuroscientists and philosophers are mesmerized by the Cartesian sacred chant — begin with the parts, begin with the parts — their overall way of understanding how the mind works is predictable: The mind is a collection of modules designed by natural selection to solve the problems faced by our ancestors in their foraging way of life on the African savannah; each module is an expert at solving one kind of problem, and, of course, a module's basic logic is specified by the human genome. This reductionistic scheme demands that all causation is upward; mental and brain functions are determined by physiology, neurons, and neurotransmitters, which in turn are caused by chemical and electrical activity within and between cells. Sperry summarizes the reductionistic outlook in neuroscience: "We and all our thoughts, behavior, and decisions, as well as everything around us, are controlled from below by strictly physico-chemical forces that reduce ultimately to quantum physics. Everything in the brain and elsewhere is subject to the laws of physics and chemistry. There is no freedom, no choice, no values, no intention, no moral priority."[173]

Sperry gives personal testimony to how free choice operates in science: "As a scientist, my world outlook with regard to both human and nonhuman nature underwent a major conversion during the mid-1960s. . . . [W]ithout abandoning or compromising scientific principles, I have come around almost full circle today to reject the type of truth science traditionally has stood for, along with its dominant central tenet that everything in our universe, including the human psyche, can be accounted for in terms entirely physical . . ."[174]

Brain Damage and the Nonmaterial

Reductionism allows for only upward causation. With this view of nature, emotional depression *must* be caused by chemical imbalances in the brain; however, common experience tells us some cases of depression, if not most, result from sad events in a person's life. Although the number of receptor molecules for the neurotransmitter serotonin present on the membranes of a certain class of blood cells is below normal in many emotionally depressed persons, depression is not always the result of brain disorder. Downward causation can also produce fewer numbers of serotonin receptors. A friend of mine Barry became severely depressed when his mother was in the hospital dying of terminal cancer. He complained to his psychiatrist that the anti-depression drugs were of no avail. "What did you expect," his analyst told him. "Your mother is still dying." My friend's depression did not result from brain malfunction.

Two years after his mother died, Barry fell madly in love with Anya. She shared Barry's passion for Russian literature, surpassed his high intelligence, and wrote better prose than he. Quick-witted, Anya was a joy to be around. Barry's whole world was transformed with Anya at the center; even his walk became buoyant, and his once gloomy vision of his future became filled with hope.

The upward causation of reductionism invokes the chemistry of love. Barry's brain was flooded with dopamine which elevated his mood to euphoria at times. A decrease in frontal cortex activity made him overlook Anya's laziness and obsession with other people's opinions. An increase in oxytocin and vasopressin induced bonding behavior; as a result, Barry married Anya. Downward causation, the holding of hands, the reading of love poetry together, and cuddling together produce an increase pf dopamine, oxytocin, and vasopressin. Barry's love for Anya did not result from brain chemistry but from meeting a woman perfect to share his life with.

No one doubts that a brain tumor or a stroke can cause radical, bizarre changes in a person's perception, thought processes, or behavior. A former student of mine, Daren, suffered a stroke in the left occipital lobe of his brain, where visual processing takes place. Before his stroke, Daren, a Canadian and a member of the First Nation, loved words and wrote beautiful prose. Post-stroke, he had difficulty finding words because he could not recognize what he saw. Daren used photographs to help work his way back to full language. Every face that he saw head-on, either in a photograph or in actual life, appeared to him as split in half, and if the left and right sides were not perfectly symmetric, he felt pain and had to turn away. He saw eyebrows, corners of eyes, and foreheads with a heightened intensity and clarity. Oddly, he never experienced the faces of dogs or cats as split apart. In addition, his central vision had shrunk; to see an entire object or person, he had to force his eyes to scan whatever was in front of him, and even then, he often could not recognize the object or person. Through cognitive therapy and determination, Daren, four years post-stroke, is vastly improved, though not his old self, yet.

Brain function reveals precious little about human life. Except for disease, such as a cerebral hemorrhage or a brain tumor, the owner of a brain does not care about neurons and neurotransmitters. Similarly, the user of Microsoft Word is not interested in logic gates or operating systems and ignores the hardware as well as the binary code of the software unless the hard drive fails or the blue screen of death appears. The inner world of perception, thought, feeling, and desire is illuminated by literature, music, and philosophy, and to a lesser extent by doing mathematics and science. The claim that science offers the sole explication of human life is bogus. How the brain functions is an interesting scientific problem, but ultimately neuroscience has virtually nothing to say about love and hate, hope and despair, freedom and enslavement.

As Schrödinger did years before me, I was amazed to discover that the scientific picture of the real world and real human beings is extraordinary deficient. Science is, in the words of Schrödinger, "ghastly silent about all and sundry that is really near to our heart, that really matters to us. It cannot tell us a word about red and blue, bitter and sweet, physical pain and physical delight; it knows nothing of beautiful and ugly, good or bad . . . . Science sometimes pretends to answer questions in these domains, but the answers are very often so silly that we are not inclined to take them seriously."[175]

Despite my gypsy heritage and my outsider life, I was seduced at an early age by the picture science paints of the world. The beauty of relativity and quantum physics made me a true believer. Unless a person is born with the quirky ability to do higher mathematics, it is almost impossible to understand that seeing the beauty of theoretical physics is akin to a mystical experience. At twenty-one, I was convinced that science was the only path to truth and scoffed at the several people I knew who were studying philosophy and made fun of the two physics graduate students who attended church regularly. Like most scientists, I believed religion belonged to the Dark Ages that preceded the Age of Enlightenment. Men and women could only be truly free when they cast off the chains of the past. I thought I had the great advantage of not having a religious upbringing, so I was free from the ridiculous ideas instilled by religious authorities.

Later in life, I discovered that all human beings suffer from a fatal intellectual flaw, the propensity to take one truth and make it the only truth. For most scientists, science is a new religion, and unlike Schrödinger, they refuse, like all true believers, to see that the emperor has no clothes, close their eyes to the limitations of the experimental method and willfully deny that the philosophy of materialism they embrace gives "silly" answers to the most fundamental questions about human life. For me, and I suspected for a growing number of young people, I had to divest myself of the "silly" ideas instilled in me by science, not religion; in the twentieth-first century, science, not the Church, is the oppressor that champions a worldview that has to be cast off.

In my quest for the meaning of human life, I soon grasped that science is only one path to truth, to truths that are not the most interesting part of life, although they are beautiful and surprising. Among J. Robert Oppenheimer's last written words were "science is not everything, but science is very beautiful."[176] Physics and chemistry examine the piping, the infrastructure of life, not the ultimate reality of the universe or human life. To hold that the ultimate reality of a performance of Suite No. 1 of Johann Sebastian Bach's *Six Unaccompanied Cello Suites* is the scraping of horsehair on catgut is a philosophical absurdity, not to mention human insanity.

The physical world provides an intense, rich interior life, which includes the impersonal, stark beauty of mathematics and physics as well as the pungency of Stilton cheese, the softness of cashmere, the dance of cherry blossoms, the smell of the ocean salt air, the wonder and mystery of nature, and the poetry, drama, and music that touch the transcendent.

11 Quantum Physics and Mind

In January 1926, Erwin Schrödinger published in *Annalen der Physik* the paper "Quantisierung als Eigenwertproblem" (Quantization as an Eigenvalue Problem) that announced what is now known as the Schrödinger equation. This paper, universally acknowledged as one of the most important intellectual achievements of the twentieth century, created a revolution in physics. Quantum physics proved to be all-encompassing: For the first time, physicists understood the structure of the periodic table of chemical elements, the conduction of electricity in metals, the energy production in stars, and the first three minutes of the universe. Surprisingly, a new technology — CD players and smartphones, for instance — emerged from quantum physics, which in 1926 seemed like an esoteric branch of physics only of interest to specialists. Despite these successes, the philosophical meaning of the quantum physics revolution is still intensely debated. In January 2017, ninety-one years after the appearance of Schrödinger's paper, Steven Weinberg, considered by many physicists as the most distinguished living member of their profession, published the essay "The Trouble with Quantum Mechanics."[177] To understand why the meaning of quantum mechanics is still controversial, we must first see the failure of Newtonian physics to describe the atomic word.

The Abandonment of Newtonian Thinking

The success of understanding nature in terms of mechanical models led nineteenth-century physicists, chemists, and biologists into the error of envisioning the future as a linear projection of the past. Many physicists thought their field was essentially complete. Addressing the *British Association for the Advancement of Science*, in 1900, Lord Kelvin, the greatest living English physicist at that time, predicted, "There is nothing new to be discovered in physics now. All that remains is more and more precise measurement."[178] Albert Michelson of the famed Michelson-Morley experiment concurred: "The grand underlying principles have been firmly established . . . further truths of physics are to be looked for in the sixth place of decimals."[179] Philipp von Jolly, an experimental physicist, advised one of his students, Max Planck, not to go into physics: "In this field, almost everything is already discovered, and all that remains is to fill a few unimportant holes."[180]

A short time later, in 1909, in Manchester, England, Hans Geiger and Ernest Marsden under the direction of Ernest Rutherford carried out the Gold Foil Experiment to probe the structure of the atom. Rutherford expected to confirm the plum pudding model of the atom proposed by J. J. Thompson, who discovered the electron in 1897. Thompson pictured an atom where electrons (negatively-charged "plums") move freely within a cloud of electrically positive substance (the "pudding"). Geiger and Marsden directed a beam of alpha particles generated by the radioactive decay of radium at a sheet of very thin gold foil. If the plum pudding model were correct, then the alpha particles would be deflected by at most a few degrees. Geiger and Marsden, however, observed that some of the alpha particles were deflected through angles much larger than 90 degrees. Rutherford was astounded: "It was quite the most incredible event that has ever happened to me in my life. It was almost as incredible as if you fired a 15-inch shell at a piece of tissue paper and it came back and hit you."[181]

Rutherford proposed, in 1911, the planetary model of the atom, where electrons (the "planets") orbit a positive charge concentrated in a tiny volume (the "Sun"). In this model, the atom is mostly empty space. Shortly thereafter, theoretical physicists using Newtonian physics and Maxwellian electrodynamics calculated, to their dismay, that electrons continuously accelerating around the nucleus of an atom would, within a hundred-millionth of a second, lose all their energy, spiral into the nucleus, and destroy the atom. If this were true, the universe would disappear in a flash. Something was terribly wrong with the mechanical universe.

Another striking failure, labeled "the ultraviolet catastrophe," was the prediction on Newtonian principles by Lord Rayleigh and James Jeans that a hot black body with a small hole through which electromagnetic energy could escape would produce an *infinite* quantity of energy — an absurd conclusion in direct contradiction to the law of the conservation of energy.

Physicist Max Planck, in 1900, after years of wrestling to understand black-body radiation, laid the foundation of quantum physics when he proposed that electromagnetic energy could be emitted and absorbed in only discrete units. With one simple equation, $E = hf$, where E is the energy element, f the frequency of the radiation, and h Planck's action constant, he precipitated a revolution that overthrew the Newtonian understanding of matter and forged a new branch of physics — quantum mechanics.

Today, the philosophical interpretation of quantum mechanics rests on the noncontroversial but mind-boggling, double-slit experiment.

The Double-Slit Experiment

Physicist Richard Feynman taught that the double-slit experiment with single electrons is "impossible, *absolutely* impossible, to explain" in any way with Newtonian physics, for in it is the "heart" of quantum physics.[182] Like most physicists in the 1960s, Feynman believed the double-slit experiment was a *gedanken* (thought) experiment because the "apparatus would have to be made on an impossibly small scale to show" quantum effects.[183] The experiment, however, had already been performed in 1961 by Claus Jönsson at the University of Tubingen and in 1974 at the University of Milan by researchers led by Pier Giorgio Merli and later in 1989 with great precision and elegance by Akira Tonomura and his colleagues at Hitachi in Japan.

In the Japanese double-slit experiment[184], electrons were emitted one by one from an electron microscope. They were fired in the direction of two vertical slits at the rate of 10 per second, each electron traveling with a velocity of 120,000 km/second so that no electron could possibly interact with any other. Figure 11.1 is a schematic representation of the Japanese experiment.

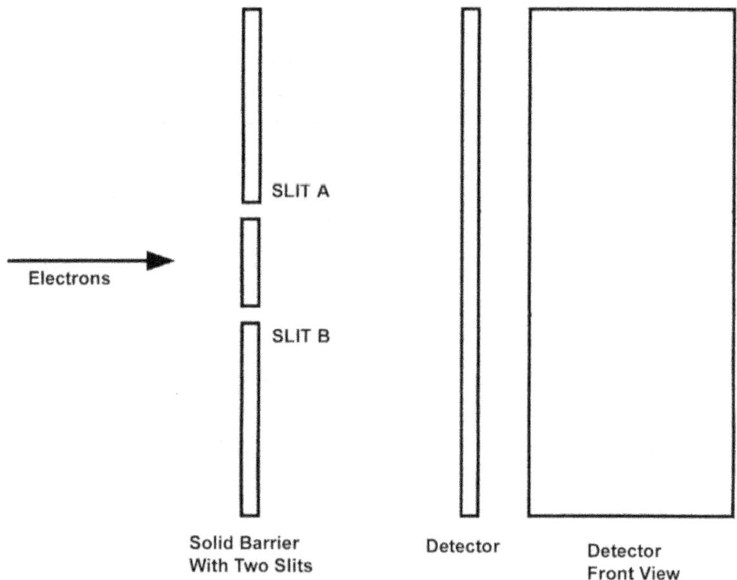

Figure. 11.1. A schematic representation of the Japanese double-slit experiment. Single electrons are fired one by one in the direction of two slits, labeled A and B.

If an electron were a Newtonian particle, something like a tiny baseball or a bullet, it would necessarily follow a well-defined trajectory. Arriving at the detector, an electron would have gone through either slit A or slit B. Electrons, then, would cluster on the detector along two vertical lines behind each slit. The lines would be fuzzy, not sharp because electrons would leave the emitter in various directions, and some would ricochet off the sides of the slit openings. The distribution of electrons at the detector would be the sum of those electrons that passed through slit A and those that passed through slit B. (See Figure 11.2.)

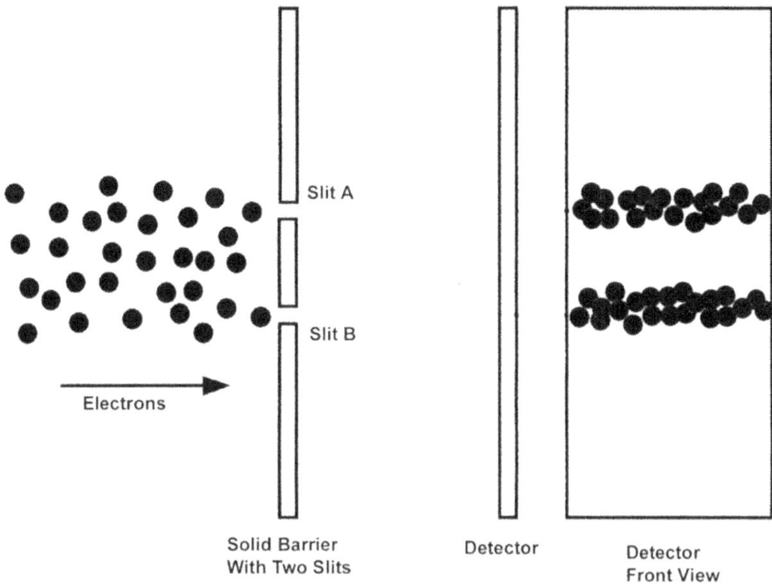

Figure 11.2. Prediction: If electrons are Newtonian particles, in the double-slit experiment, they would cluster on the detector behind slit A and slit B.

The results of the experiment were most unexpected and completely contradicted the Newtonian prediction. (See Figure 11.3.) The number of electron hits was maximal *directly behind the barrier* separating the two slits! The electrons that hit the detector formed a symmetric pattern around this maximum. To the immediate sides of this maximum were two minima, places where fewer electrons struck the detector. Along the detector, a series of alternating maxima and minima occurred, with each maxima less than the preceding one. (The alternating maxima and minima of the distribution of electrons on the detector are called interference fringes because they resemble the decorative fringes sometimes used on clothing and rugs.)

A physicist cannot say that an electron passed through either slit A or slit B, for if that were true, then the Newtonian pattern in Figure 11.2 would have resulted in the experiment. Unlike the bodies of Newtonian physics, where the three laws of motion keep a planet on its orbit or direct a bullet toward its target, an electron does not follow a well-defined trajectory. In the double-slit experiment, a physicist cannot say where an electron is until it is observed at the detector.

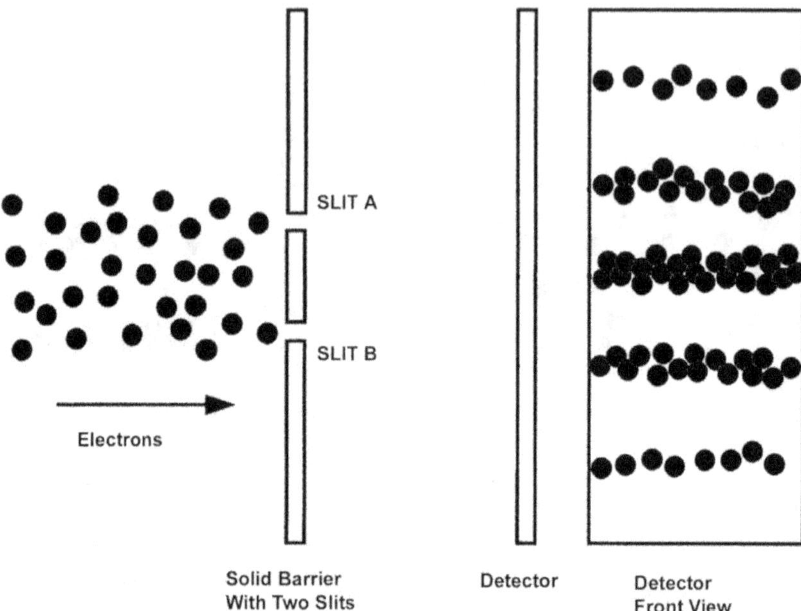

Figure 11.3. A schematic representation of the results of the double-slit experiment with single electrons. The maximum number of electrons on the detector occurs directly behind the barrier separating the two slits. See text for a description of the symmetrical series of alternating maxima and minima.

The Japanese experimenters watched the pattern of hits on the detector slowly build up over time. At first, the electrons appeared as random hits on the detector plane (see Figure 11.4a). As electrons accumulated, a coherent pattern seemed to form. Finally, in Figure 11.4d, a pattern was clearly visible. Contrary to the expectation of Newtonian physics, the electrons did *not* cluster around two lines. Instead, one by one, the electrons over time produced a pattern of alternating maxima and minima.

We have reached the profound mystery of quantum physics. How is an alternating pattern of maxima and minima produced by electrons that arrive at the detector in an unpredictable manner? No one has any idea how or why this happens; physicists can just describe the results. All the fundamental equations of quantum physics — the Schrödinger and the Dirac equations among others — do not describe the behavior of individual atoms, electrons, photons, or other subatomic particles. The Schrödinger equation has nothing to say about Figure 11.4a and only speaks about Figure 11.4d. The individual electrons in the double-slit experiment arrived at the detector in a completely unpredictable manner. (See Figure 11.4a.) How nature goes from Figure

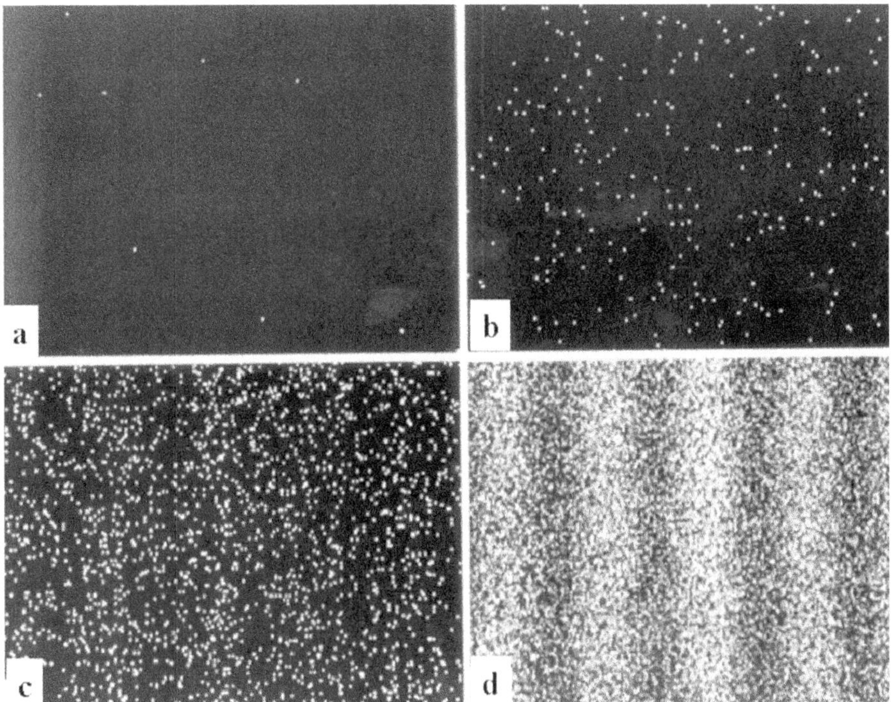

Figure 11.4. The actual data from the Japanese double-slit experiment. Single random electron events build up to form a coherent pattern. See the remarkable video clip on the Hitachi website produced by the experimenters from the data.

11.4a to Figure 11.4d is probably a permanent mystery. According to Feynman, physicists "now believe it is impossible"[185] to predict individual events, such as those seen in Figure 11.4a.

We should point out a tempting error — the belief that an electron in reality must go through either slit A or slit B; the experimenter just does not know which one. But if an electron *actually* goes through either slit A or Slit B, then detectors at the slits can determine which slit the electron goes through and will *not* affect the outcome of the experiment. But it can be shown that such detectors destroy the interference pattern and produce what is predicted by Newtonian physics (Figure 11.2). Thus, we must conclude that each electron in the experiment has the potential to go through either slit, and this potentiality is the cause of the interference pattern. *An elementary particle exists in a state of potentiality until observed.*

In the double-slit experiment, no cause can be given why two electrons struck the detectors at different places. Individual events are unintelligible; only a large number of events, what physicists call an ensemble, exhibit mathematical regularity.[186] Stated another way, individual events are irrational, for they fall outside of reason; only a large number of events exhibit statistical regularity. The universe is not deterministic as promulgated by the worldview of Newtonian physics.

One firm conclusion of quantum physics is that all elementary particles and atoms, exist as potentialities or possibilities rather than as ordinary objects like billiard balls. Werner Heisenberg notes that atoms and elementary particles are not fully actual but "form a world of potentialities or possibilities rather than one of things or facts."[187] The assumption that electrons possess full existence like tiny, steel BBs contradicts the double-slit experiment. An electron, of course, is not entirely indeterminate in every respect. An electron's mass is $9.10938188 \times 10^{-31}$ kilograms, and its charge is $1.60217646 \times 10^{-19}$ coulombs. These two properties are universal constants in every branch of physics and chemistry.

Atoms and elementary particles become actual only when an experimenter makes an observation. If an experimenter measures the location of an electron, she finds the electron at a particular place. If the experimenter measures the electron's velocity, she determines its speed. But in accord with the Heisenberg Uncertainty Principle, the experimenter cannot simultaneously determine both the location and the velocity of an electron. What potentialities of the electron are actualized depends upon the observations of the physicist and thus on her choice of measurement strategy. Here the term "observation" always entails actualizing some aspect of the particle. Heisenberg describes how the physicist and elementary particles are related: "We can no longer talk of the behavior of the particle apart from the process of observation . . . the laws of nature which we formulate mathematically in quantum theory deal no longer with the particles themselves but with our knowledge of the elementary particles."[188]

Renegade physicist David Bohm sums up the essential feature of quantum physics: "The primary emphasis is now on *undivided wholeness*, in which the observing instrument is not separated from what is observed."[189]

Scientific Realism

That elementary particles exist in potentiality until actualized through measurement is in direct opposition to scientific realism, the prevailing outlook of modern science from its very beginning. Galileo, in 1623, argued that tastes, odors, and colors reside only in human consciousness, and all these qualities would be wiped away and annihilated "if the living creature were removed."[190] The senses do not report reality; "the office of the sense shall be only to judge of the experiment, and the experiment itself shall judge of the thing."[191] (See Figure 11.5.) Reality is what is left behind when the human creature is removed. The goal of science, according to this view, is to understand nature in the absence of the scientist.

Figure 11.5. The Tevatron at Fermilab, near Chicago, Illinois, judges the top quark, and physicists judge the output of the Tevatron. The experiment touches nature, and the scientist touches the experiment. (The Tevatron went permanently dark in September 2011. Photograph courtesy of Fermilab.)

Isaac Newton added the final two elements of scientific realism when he published, in 1687, *Philosophiæ Naturalis Principia Mathematica*: 1) the universe is mechanical and 2) every whole is completely understandable in terms of its smallest parts and how they interact. For the next 300 years or so, physicists, biologists, and neuroscientists labored to show that "the universe, including all aspects of human life, is the result of the interactions of little bits of matter."[192]

The Revolution

In the double-slit, the elementary particles, the measuring apparatus, and the experimenter form an undivided whole. Unlike scientific realism, in quantum mechanics the experimenter is a participant in nature as well as an observer; an understanding expressed by Bohr: "In the great drama of existence, we ourselves are both actors and spectators."[193] Physicists have expressed this fundamental aspect of quantum mechanics in various ways. Eugene Wigner: "It was not possible to formulate the laws of quantum mechanics in a fully consistent way without reference to the consciousness."[194] Max Born: "No description of any natural phenomenon in the atomic domain is possible without referring to the observer, not only to his velocity as in relativity, but to all his activities in performing the observation, setting up instruments, and so on."[195] Freeman Dyson: "The laws of subatomic physics cannot even be formulated without some reference to the observer. The laws leave a place for mind in the description of every molecule."[196]

The Rearguard

If a recent poll of 33 attendees of a conference on the foundations of quantum mechanics[197] is representative of all physicists, philosophers, and mathematicians, then the rearguard of the Revolution outnumbers the revolutionaries on the front line storming the citadel of scientific realism, shouting "Truth is error burnt up." Fewer than two out of five physicists hold the interpretation of quantum mechanics presented here, essentially the Copenhagen Interpretation with neo-classical holdovers, such as "wave-particle duality" and "the collapse of the wave function," stripped out. The most counterrevolutionary position adheres to the advice given to graduate physics students — "Shut up and calculate!"[198] — be mindless, do not fret over the meaning of quantum mechanics, which is philosophy, mere words.[199]

Weinberg objects to what he labels the instrumentalist interpretation of quantum mechanics (more or less the one given here) on the grounds of scientific realism. He maintains that the aim of science is "to say what is really going on out there." By dragging humans into the fundamental laws of nature, the instrumentalist abandons the realist vision that the "world [is] governed by impersonal physical laws that control human behavior along with everything else."[200] Yet, Weinberg does not direct us to those impersonal physical laws that control everything and that also explain quantum phenomena, for no such laws exist; the universe is not deterministic.[201]

Currently, there are two Big Pictures of the cosmos: 1) The universe is made of parts; humans, like kangaroos and zebras, are not essential parts, only quarks and leptons are; the universe, including all aspects of human life, is the result of the interactions of little bits of matter; and 2) Things exist only in relationship and the main feature of the universe is undivided wholeness; humans, the only rational creatures we know of, are an essential element of the universe, and thus, the universe is composed of two constituents — mind and matter — neither reducible to the other.[202]

Surprisingly, the meaning of quantum physics, a most arcane subject, remote from human experience, raises the most profound question a person can ask — "Who am I?" Scientific realism gives one answer: "'You,' your joys and your sorrows, your memories and your ambitions, your sense of personal identity and free will, are in fact no more than the behavior of a vast assembly of nerve cells and their associated molecules.[203] You're nothing but a pack of neurons.[204] Genes created you, body and mind; their preservation is the ultimate rationale for your existence.[205] Your desire for happiness is at loggerheads with the whole world; that you should be happy is not included in the plan of 'Creation'.[206] You are the farcical outcome of a chain of accidents.[207] The human race is a mere chemical scum on a moderate-sized planet,[208] orbiting around a very average star in the outer suburb of one among hundreds of billions of galaxies. You are a speck of fleeting dust in an indifferent universe."

Scientific holism inspired by quantum mechanics answers "Who am I?" in a contrary way: "You are body and mind. The causal potency of an idea or an ideal is just as real as a molecule, a cell, or a nerve impulse.[209] Mind is the primary element of the universe. What possible sense could it make to speak of 'the universe' unless there was someone around to be aware of it?[210] You have the capacity to be connected to all that is.[211] Without you and other human beings, the universe would be a drama played before an empty theater and thus would be pointless.[212] Remove humankind from nature and you erase the perception of all its wonder, its beauty, and its mystery — the world becomes meaningless."

What a surprise. The founders of the citadel of scientific realism, true revolutionaries in their day, spawned counterrevolutionaries, defenders of the modern ideology, the philosophy of materialism. Over ninety years have passed since the publication of the Schrödinger equation, so we can safely assume that the counterrevolutionaries are pursuing the blind hope that someday, somewhere, some brilliant physicist will understand quantum phenomena in terms of laws that control everything — a return to the deterministic outlook of Newtonian physics.

Those partisans allied to the Copenhagen Interpretation of quantum mechanics are today's true revolutionaries, locking arms with Greek philosophers and Christian theologians, the traditional guardians of human freedom and dignity.

12 A New Technology, A New World

Francis Bacon's great insight, "those twin objects, human knowledge and human power, do really meet in one,"[213] led to a phase change in history: The "great mass of inventions"[214] that flowed from experimental science transformed how humans live together.

Mechanical Technology

During the nineteenth century in Western Europe, the train was the symbol of technological progress. The first railroad joined Brussels to Paris, a 200-mile trip that took twelve hours, a quarter of the time of that by stagecoach. Later, trains reach the speed of fifty miles per hour, causing many passengers to both marvel and take fright.[215]

Railways shrunk distances and compressed time. The hinterlands were brought closer to the urban centers. Poet Heinrich Heine, in 1845, called the train a "providential event," comparable to the invention of gunpowder and the printing press, "which swing mankind in a new direction, and changes the color and shape of life." With respect to the new experience of space, he said, "I feel as if the mountains and forests of all countries are advancing on Paris. Even now, I can smell the German linden trees; the North Sea breakers are rolling against my door."[216] All of Europe seemed to be in motion; the poet Baudelaire defined Modernity as "the transient, the fleeting, the contingent."[217] Karl Marx analyzed how the "annihilation of space and time"[218] by railways and the telegraph globalized commerce.

People were on the move, too. The travel agency Thomas Cook and Son, in 1851, sold 165,000 return tickets on special excursion trains to the Great Exhibition in London. Cook and Son later organized packages tours to Switzerland and Italy.

The world became much larger culturally, too. Trains reduced shipping costs that led to the growth of bookstores in provincial towns. Stations along the well-traveled rail lines had lending libraries and bookstalls. Travel on the rails fueled a boom in popular fiction, for many passengers desired entertaining literature to pass the time. Trains brought a provincial public to the museums, concert halls, and theaters in the cities and traveling artistic groups to the provinces.

In twentieth-century America, the automobile, not the train, became the symbol of freedom. The title of Jack Kerouac's novel *On the Road* captured the desire buried deep in every American heart — to be free, to throw off all social constraints. For many, the suburbs were an escape from the confines of a narrow ethnic community in an unglamorous part of the inner city. The automobile separated families and eventually killed the extended family in America.[219]

The Division of Labor

Automobiles, trains, and all mass technology rest upon a new means of material production — the division of labor. Instead of defining the division of labor, Adam Smith, the founder of modern economic theory, gives in his *Wealth of Nations* the example of the manufacture of pins. An unskilled workman could perhaps make twenty pins a day. With the advent of industrialism, the task of making a pin is broken into eighteen simple operations, with each workman skilled in performing one or two simple steps. Ten semi-skilled workers "could make among them upwards of forty-eight thousand pins in a day."[220] If each worker completed all the tasks by himself, then the ten workers at best could make only two hundred pins. The division of labor increased production by 240-fold! No wonder Smith extols the division of labor as the greatest innovation in material production, ever.

Smith predicted that the vast number of material goods produced through the division of labor would require large markets, or in modern parlance a consumer society, where "a general plenty diffuses itself through all the different ranks of the society."[221] No one can doubt that two hundred years of capitalism in America and Europe created for the wealthy and the poor a superabundance of goods. The typical Walmart Supercenter carries 142,000 different items. A shopper at Kroger's or Whole Foods can buy blueberries in December grown in Chile, fresh flowers flown in from Columbia, and organic lamb imported from New Zealand. Computers, TVs, and cell phones are everywhere, in the ghetto as well as aboard yachts.

To continue to exist, the capitalist economy must constantly produce new consumer goods, not unlike a shark that must keep swimming or die. Without the "great mass of inventions" that flowed from science and technology, capitalism would have ground to a halt once markets were saturated by an abundance of goods. For instance, without John Bardeen, Walter Brattain, and William Shockley, the inventors of the transistor, the iPhone would not exist, except for Dick Tracy, the comic-book character with seemingly impossible gadgets.

New inventions and technologies continually produce new goods and thus previously unknown desires, and as a result, we are all placed on the treadmill of desiring more and more. The marketplace instills values: For "a general plenty" to be consumed, an unlimited desire for material goods must be developed through advertising and mass media. In this way, everyone's attention is focused on the good life in this world, away from salvation, at the expense of intellectual and spiritual values. Every increase in the wealth of a consumer society goes into more material goods and higher profits, not into the development of the worker-consumer as a human person.

The dark side of capitalism is that the division of labor contracts the interior life of the worker. In his movie *Modern Times* (1936), Charlie Chaplin brilliantly portrays the effect of performing one repeated task in a manufacturing process upon a person. Under the opening credits, the hands on a clock face approach 6 o'clock. In the style of a silent movie, a title give the film's theme: "*Modern Times.* A story of industry, of individual enterprise — humanity crusading in the pursuit of happiness." The film opens with an overhead shot of sheep jostling in a pen, and then rushing through a chute, presumably off to the slaughterhouse. The sheep dissolve into industrial workers shouldering each other as they rush out of a subway station on their way to work.

In an early scene, the Tramp (Charlie Chaplin) is an assembly-line worker who tightens bolts on machine parts as they pass by him with clock-like regularity on a conveyer belt. His tasks require mechanical, repetitive motions. During a short break, the Tramp cannot stop his arms from continuing the jerky, rhythmic movements of nut-tightening.

Figure 12.1. A still from *Modern Times.* The tramp (Charlie Chaplin) makes a valiant attempt to keep up with the assembly line and is dragged into the machinery. Courtesy of Wikimedia Commons.

In the late afternoon, the boss orders an increase in production. The conveyer belt speeds up, and the Tramp makes a valiant attempt to keep up with the machinery, but it is impossible. He goes berserk and lies on the conveyer belt to tighten the nuts of the pieces he missed as they went by him. Dragged into the machinery, his body stops the gears and cogs of the monstrous machine (see Figure 12.1). The production line reverses, and the Tramp emerges in a trance, still with his two wrenches, now raised high. Everywhere he sees nuts that need to be tightened. He even chases two women whose suit buttons resemble the nuts he had been tightening. The interior life of the Tramp collapsed to what served the machine.

Surprisingly, in the concluding part of *The Wealth of Nations (1776)*, Smith anticipates Chaplin's devastating critique of capitalism. Smith argues that when a person performs one or two simple operations as required by the division of labor, he "generally becomes as stupid and ignorant as it is possible for a human creature to become. The torpor of his mind renders him not only incapable of relishing or bearing a part in any rational conversation, but of conceiving any generous, noble, or tender sentiment."[222] The interior life of an industrial worker collapses to what serves the machine. Smith maintains that industrialism produces an abundance of goods *and* a decline in the worker's interior life. Alexis de Tocqueville agrees: "As the principle of division of labor is ever more completely applied, the workman becomes weaker, more limited, and more dependent. The craft improves, the craftsman slips back."[223]

In premodern economies, Smith notes, no one fell into the "drowsy stupidity" induced by the division of labor and "every man ha[d] a considerable degree of knowledge, ingenuity, and invention," because the labor of the artisan developed the whole person.[224]

Neither Smith nor Tocqueville suspected that the division of labor adversely affected the physical health of the worker. The British statistician William Farr, in 1843, compared life expectancies in three parts of England: Rural Surrey residents enjoyed a life expectancy of close to 50 years; metropolitan Londoners came in at 35; the average worker in industrialized Liverpool died at 25.[225]

Digital Technology

Once again, human history is undergoing a phase change caused by a new science. Stated succinctly, the fundamental principle of Modernity, *things exist in isolation, as separate entities,* is being replaced by *undivided wholeness* of quantum physics. The experimenter and the observing instrument in quantum physics are not separate from what is observed. The knower and the known form an indivisible whole; to speak of the universe in the absence of any knower is absurd. Niels Bohr, the guiding light of the Copenhagen Interpretation of quantum physics, states the new understanding of the scientist, "In the great drama of existence, we ourselves are both actors and spectators."[226] The isolated, autonomous individual is a cultural lie.

The early stages of historical phase changes are always confusing, difficult to discern clearly, and resisted by many; consequently, the current phase change is easiest to see in the new digital technology, just as the previous phase change became immediately apparent to poets who grasped how trains were transforming economic and social life in Europe.

The Whole World

The transistor, the personal computer, and the cell phone were made possible because of the deep understanding of matter provided by quantum physics. The Apple II went on sale in June 1977 and the IBM PC in August 1981. Tim Berners-Lee, while a fellow at CERN in 1989, invented the World Wide Web. The mouse and the Graphical User Interface made the internet available for the technically ignorant; Netscape, the first browser, was released on December 14, 1994. The first smartphone, iPhone 1, was released on June 29, 2007. According to Statista, there are 3.3 billion smartphone users in the world today. Eighty-two percent of Americans use a smartphone.[227] Go into any coffee shop, college dining hall, or airport lounge, and it seems like everyone is on a smartphone, texting a friend, reading email, surfing the Web, or watching a movie.

The ubiquity of digital devices makes the number of people who share the same experience staggering. Take YouTube. The views of *Leonard Cohen: Hallelujah (Live in London)* is 147 thousand, of *Justin Bieber: Sorry* 3.2 billion, and of the children's cartoon *Masha and The Bear: Recipe for Disaster*, originally in Russian, 4.2 billion. The comparison number here is the human population of the Earth, 7.7 billion.

Notre Dame Cathedral in Paris catches fire, and local videos of the catastrophic blaze are immediately on social networks. A volcano erupts on White Island, New Zealand, killing seventeen tourists, and within hours, videos of other tourists being rescued and of the Prime Minister of New Zealand, Jacinda Ardern, giving an update of the tragic event are seen in Peoria, Illinois.

The previous cultural shift from the provincial to the urban has been expanded to the entire world by digital technology, and that shift has political consequences.

Our Choices Affect the Whole World

In 2015, Christine Figgener, a marine biologist, found a male olive ridley sea turtle during a research trip to Costa Rica. The turtle had a four-inch plastic straw lodged in its nostril. Dr. Figgener videographed herself using a pair of pliers to remove the straw that caused great pain and substantial bleeding to the turtle. Most of us have not lived on a farm and seen chickens or sheep slaughtered for our dinner, so we cringe when we see a poor turtle in pain and bleeding, even though it is being rescued.

Dr. Figgener's YouTube video[228] went viral with over 38 million views and gave the movement to ban plastic straws a tremendous boost. In Postmodernity, the political consequences of viral videos can develop rapidly. July 2018, Seattle became the largest U.S. city to ban plastic straws. January 2019, California became the first state to ban plastic straws at full-service restaurants. By 2020, Starbucks planned to ban plastic straws from all its stores.

In the previous era of Newtonian science, in the Mechanical Age, the efficiency of production, the convenience to the consumer, and the creative destruction of capitalism were the principal "moral" values. In the Digital Age, the nearly universal moral question is "How does our behavior affect other persons, animals, and the environment?" This is not a question just asked by bleeding-heart liberals.

Tucker Carlson on his weekday program on Fox News featured a story about Paul Singer, a hedge fund manager and the second largest contributor to the Republican Party in 2016. Elliot Management, Singer's hedge fund, bought eleven percent of Cabela's, a large retailer that sells fishing and hunting gear. To make a quick profit on his investment, Singer forced the management of Cabela's to sell their company to Bass Pro Shops, even though Cabela's turned a profit on its $4 billion of total revenue the previous year. The new owner moved the headquarters of Cabela's with its 2,000 good-paying jobs from Sydney, Nebraska, population 6,282. Sydney was destroyed, but Paul Singer made at least $93 million.

Tucker Carlson could not hide the moral outrage in his voice when he said the model for the destruction of America is "ruthless economic efficiency. Buy this distressed company, outsource the jobs, liquidate the valuable assets, fire middle management, and once the smoke is cleared, dump what remains to the highest bidder, often in Asia."[229])

The Sophoclean Curse

In Chapter 1, we suggested that the change from one historical era to another is a phase change like the melting of ice into water, each substance with its own properties, but unlike nature, the phase changes of history cannot be reversed. The other feature of historical change was enunciated by Sophocles 2,500 years ago: "Nothing that is vast enters into the life of mortals without a curse,"[230] which we have already seen with the introduction of the division of labor.

After World War II, mass immigration from the Third World and Asia brought white, middle-class Americans into personal contact with "strange" food, music, and traditions. Salsa, now, outsells ketchup; reggae outlasted Lawrence Welk; and spiritual practice means Eastern to more and more people. Despite ourselves, we are being freed from cultural prejudices by communications technology, the global marketplace, and shifting populations. Beliefs and practices that are purely conventional and arbitrary are being exposed as such. Deep in our hearts, we Westerners know we can no longer believe that we possess the one true way of living.

The old guard continues to pursue securing economic and political spheres of dominance; the abandonment of military conscription allows the American government to engage in military operations in Africa, Afghanistan, and the Middle East with citizens at home taking little notice. But the lives of fewer and fewer young Americans are dominated by the patriotic stories told by their flag-waving great-grandparents. The digital connection to the whole world is destroying nationalism, and consequently for many, the Nation-State has lost its divine aura. For the youth of the Digital Age, the new reality that is replacing the Nation-State worldwide is the Big Five — Amazon, Apple, Facebook, Microsoft, and Alphabet, the parent company of Google.

Social media, not nationalism, is becoming the new glue that holds Americans together: 27 percent of adult Americans use Snapchat, 35 percent Instagram, 68 percent Facebook, and 73 percent YouTube.[231] Social media is causing society to fragment into small, digitally connected groups. The median number of friends a Facebook user has is roughly 150.[232] But these small groups are not like those of small New England villages in the 1800s, where Yankees lived together daily and could not avoid face-to-face contact.

A farmer plowing his field in the Merrimack Valley in New Hampshire had memories of family, neighbors, and people he passed on the road going to and from the market and knew the names of characters in the Bible and outstanding patriots, such as Washington and Jefferson. At school, he had learned stories, ballads, and episodes of American history that taught a moral lesson. With the advent of the image world of movies, television, and the internet, everything changed. Movie critic Geoffrey O'Brien declares that in the new image culture "half the faces in the memory bank [of a person] are of public personalities, actors playing fictional characters."[233] The internet has greatly expanded the world and at the same time weakened human relationships.

According to a 2016 Nielsen report, the total amount of time American adults spend on average in the Image-World, namely watching TV, surfing the Web on a computer, and using an app on a smartphone, is eight hours and forty-seven minutes per day.[234] As a result, Americans acquire the habit of judging people and products by feelings; likes and dislikes determine whether a song, a movie, a sitcom, a celebrity, or a political candidate is up or down. In the Image-World, reasoned argument is replaced by emotional judgment.

Technology administered the *coup de grâce* to truth. The internet is the first medium in history, besides the agora of ancient Athens and the town meetings in New England villages, to create many-to-many communication; the phone is one-to-one, and books, radio, and television are one-to-many. Clay Shirky, a prominent thinker on the social and economic effects of the internet, points out that "every time a new consumer joins this media landscape a new producer joins as well, because the same equipment — phones, computers — let you consume and produce. It's as if, when you bought a book, they threw in the printing press for free."[235]

Social media allow hundreds of millions of Americans with the click of a mouse to disseminate their opinions without the scrutiny of grammarians, fact-checkers, and editors, the guardians of print culture. Anyone, anywhere, anytime, now, can instantly post an opinion on anything. The internet has become an ocean of democratic opinion, one comment washing over another, quickly submerging whatever truth that tries to surface.

Expert assessment based upon objective, rational analysis reeks of elitism. Scientific data and peer-reviewed reports on climate change are posted on the internet, but videos of fires in New South Wales, Australia and photographs of thawing permafrost and sinking tundra in Newtok, Alaska move people to protest and sign petitions against fossil fuel corporations.

The New Hope: Artificial Intelligence

History is not a predictive science like Newtonian physics; human beings have free will that allows them to behave imaginatively for good or bad; the tragedy is that they are often blind to the consequences of their actions. Reading the tea leaves of Post-Modernity, we are limited to pointing out possibilities, knowing that we may miss what turns out to be the most important emerging pattern that is obvious fifty years from now.

Computers have revived the Baconian hope among technologists that the "great mass of inventions" will lead to Paradise on Earth, that machines will take over the mindless work now done by humans, so more and more of us will be freed to improve our lives through high culture and insight meditation. Today's Baconians place their hope in deep learning, the core of artificial intelligence (A.I.). Computers programmed with deep learning take a huge amount of data within a single domain and learn to predict or decide at superhuman speed and accuracy. Many jobs are mechanical in the sense that the same operation is performed again and again. Consider the diagnosis of diabetic retinopathy, a condition where the nerves and blood vessels in the retina are damaged by complications of diabetes. Computers were trained to search for patterns in photographic images of the retinas of diabetics and non-diabetics. After several months of training, computers achieved a 95 percent accuracy in diagnosing diabetic retinopathy, an accuracy at or above the average ophthalmologist, who required thirty years of schooling and specialized training.[236]

Many jobs that we do not think of as mechanical can supposedly be replaced by A.I. computers. The self-driving car is already almost four times safer than a human-operated car: one crash every 1.92 million miles versus one crash every 492,000 miles.[237]

In a world where computers outperform us in analyzing data, carrying out mechanical operations, and increasing efficiency, we are left with what differentiates us from computers — imagination, insight, and compassion. A computer is indifferent to whether a patient is diagnosed with two months to live or is disease-free. A doctor or a nurse can hold the patient's hand; a priest or psychotherapist can help the patient confront his mortality. Modern society desperately needs caregivers for the ill and elderly, teachers for children, guides for young adults, and love for the neglected, abandoned, and forgotten.

According to present-day Baconians, A.I. can liberate us from repetitive labor that Smith said renders the worker "not only incapable of relishing or bearing a part in any rational conversation, but of conceiving any generous, noble, or tender sentiment."[238] The industrialism that required that humans perform like machines — a demand that did violence to the soul as well as the body — is destined to end. Machines, then, will do the work of machines, somewhat like the original Garden of Eden, where supposedly "the agricultural implements worked for him [Adam] of themselves, like automata."[239]

Neo-Luddites recall that the *United States Democratic Review* predicted in 1853 that within half a century "men and women will have no harassing cares, or laborious duties to fulfill. Machinery will perform all work — automata will direct them. The only tasks of the human race will be to make love, study, and be happy?"[240] Machines will never save us, only a general commitment to excellent human living will.

Be More, Not Have More

Before the full realization of Modernity, Tocqueville wrote, "In America, I have seen the freest and best educated of men in circumstances the happiest to be found in the world; yet it seemed to me that a cloud habitually hung on their brow, and they seemed serious and almost sad even in their pleasures," because they "never stop thinking of the good things they have not got."[241]

Material prosperity in America did not bring about a general happiness; from 1940–1995, as material prosperity increased, people reported a decrease in happiness, according to the *Statistical Abstract of the United States*.[242] Figure 12.2 shows that for more than fifty-some years, personal income has increased substantially, but the percentage of people who report they are very happy has not budged.

Psychologist Daniel Kahneman and economist Angus Deaton analyzed the responses of more than 450,000 United States residents surveyed in 2008 and 2009 about their emotional well-being. Kahneman and Deaton focused on an "individual's everyday experience — the frequency and intensity of experiences of joy, stress, sadness, anger, and affection that make one's life pleasant or unpleasant." They sought an answer to the question "does money buy

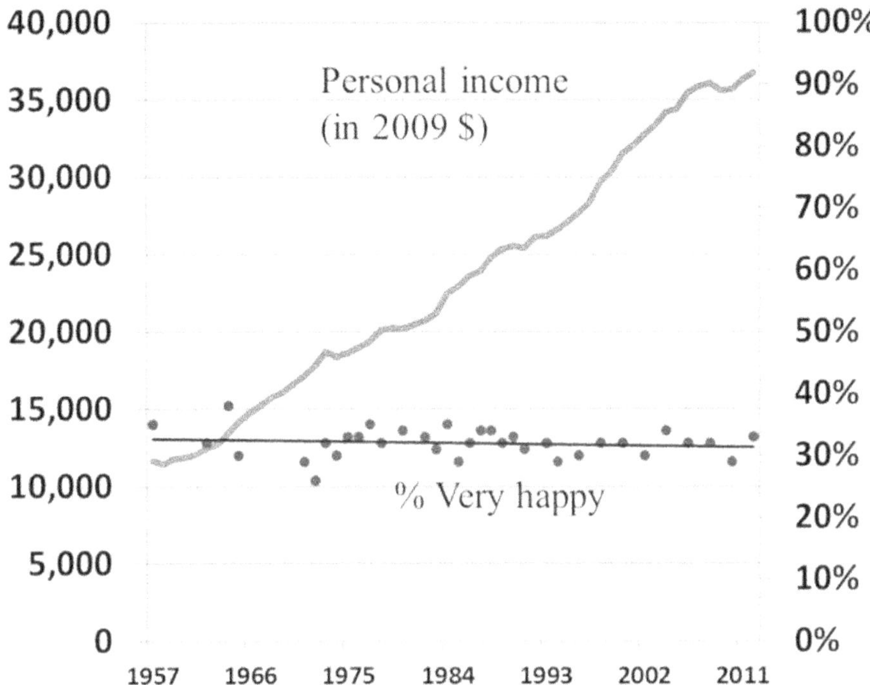

Figure 12.2. Happiness and Income in the United Sates.
Courtesy www.davidmyers.org.

happiness?" Not surprisingly, "low income exacerbates the emotional pain associated with such misfortunes as divorce, ill health, and being alone." Unexpectedly, Kahneman and Deaton discovered that emotional well-being does not increase significantly beyond an annual household income of $75,000. "More money does not necessarily buy more happiness."[243]

For nearly a century, Americans have been acquiring more and more, believing material prosperity equals happiness. Go into any Walmart, Costco, Bergdorf Goodman, Nordstrom, or Neiman Marcus — an amazing abundance of goods and so little happiness. Perhaps, just perhaps, truth has burnt up error, and we have learned that we are spiritual, not material beings. Then, the goal of material production, is to help realize the aspiration to be more, not the craving to have more.

A New World

We desperately need a vision that life is good.

In the new world struggling to be born, the madness of acquiring more and more is absent. Inhabitants desire to be more, to connect with others through love and friendship, to develop the interior life through poetry and music, to seek wisdom through philosophy and theology, and to fully embrace that they are spiritual beings and forcefully reject that they are collections of quarks and leptons, mere meat machines driven by pain and pleasure. Wide recognition of our true nature has enormous political and economic consequences.

Consider the current solution to climate change: windmills and solar farms to eliminate fossil fuels and thus reduce CO_2 emissions while *continuing* unabated our mad acquisition of more and more. The obvious solution to climate change is to alter our way of living, so we consume far less. But most of us are like alcoholics; we know our addiction is bad for us, will never give us lasting happiness, and only introduces new problems in our lives; yet, we cannot stop our destructive behavior. The latest iPhone, the cruise to Hawaii, and the new Prius are drug hits that quickly fade.

A widespread desire to be more leads to an economy where natural resource use is stabilized within ecological limits and where the goal of an increasing Gross Domestic Product is superseded by the goal of the physical, social, and spiritual well-being of everyone. The enormous wealth generated by technology can support poets, artists, and musicians as well as professional athletes, hedge fund managers, and vulture capitalists. The inevitable loss of repetitive, mechanical labor can lead to architects, engineers, and construction workers rebuilding the postindustrial landscape, where strip malls and urban expressways are a visual nightmare of the past. The threat of a war that destroys humanity through the insane use of nuclear weapons is always present, although the weakening of the Nation-State does lead to an average citizen uninterested in a National Empire and the division of the world into economic and political spheres of dominance.

The above vision is not utopian; evil, suffering, class structure, and the rich and the poor will always be with us. The twentieth century, the century of mass political murder, demonstrated that any attempt to establish Paradise on Earth ends in disaster; however, we can strive for a better society that aims to establish the well-being of all and a culture that embodies truth, beauty, and justice.

13 Paradise on Earth

The great mass of inventions that flowed forth from the new science founded on experiment, not on the Aristotelian observation of nature, captivated peasants, shopkeepers, and intellectuals alike. Everyone seemed to experience the positive emotion accompanying the belief in the dawning of a new age heralding a bright future, where all the current ills of human life — poverty, illness, and strife — would be overcome. That Bacon introduced a phrase change in human history was universally acknowledged; however, the dark side of placing human salvation in experimental science as the only path to truth was unforeseen. Blinded by the forthcoming material well-being, virtually no one saw the wisdom of Sophocles: "Nothing that is vast enters into the life of mortals without a curse."

The Second Fall

Unlike Eve eating fruit from the forbidden tree, the beginning of the Second Fall can be precisely dated: In October 1620, Francis Bacon, the principal architect of the experimental method of modern science, published *The Great Instauration*. The word "instauration," once a common English word, now sounds as if it belongs in a dreary Latin grammar book — *instauro, instaurare, instauravi, instauratus*. The Bing Dictionary gives an excellent definition of instauration: "the restoring of something that has lapsed or fallen into decay."[244] Bacon argued that what had fallen into decay was human knowledge. In the Garden of Eden, Adam named the animals according to their nature and this knowledge gave Adam command over nature.

So out of the ground the LORD God formed every beast of the field, and every bird of the air; and brought them to the man to see what he would call them; and whatsoever the man called every living creature, that was its name. The man gave names to all cattle, and to the birds of the air, and to every beast of the field.[245]

As we saw in Chapter 1, to partially restore humankind to the Garden of Eden, the only time in human history that humans had real authority over nature, Bacon declared a new science was needed "in order that the mind may exercise over the nature of things the *authority* which properly belongs to it."[246]

In establishing the philosophy of experimental science that still prevails, Bacon introduced a practical element entirely new to humankind: The true test of human knowledge is whether nature can be commanded, for "those twin objects, human knowledge and human power, do really meet in one."[247] In Bacon's grand vision of the future of humanity, the meeting of those twin objects would make humans masters and possessors of nature, much as Adam was in the Garden of Eden.

Bacon delivered two admonitions to his successors about the use of the new experimental method. First, he admonished scientists that the study of no part of nature is forbidden, "for it was not the pure and uncorrupted natural knowledge whereby Adam gave names to the creatures according to their propriety, which gave occasion to the Fall."[248] In their pursuit of the truth, scientists have a divine mandate to ignore religious dogmas, philosophical doctrines, and cultural taboos. In effect, Bacon sanctioned the separation of science from religion *and* philosophy.

Second, Bacon addressed "one general admonition to all — that they consider what are the true ends of knowledge, and that they seek it not either for pleasure of the mind, or for contention, or for superiority to others, or for profit, or fame, or power, or any of these inferior things, but for the benefit and use of life, and that they perfect and govern it with charity."[249] The general admonition exposes an inherent flaw in the *Great Instauration*. Bacon hoped to restore humankind to the Garden of Eden intellectually but knew humankind could not be restored morally.

In Bacon's grand vision, the command of nature would give to morally corrupt men and women, capable of great good and immense evil. In *Dialogues Concerning Two New Sciences*, Galileo announces his monumental discovery that terrestrial matter does indeed obey mathematical laws, and consequently, the Earth and the stars are made of the same material. At the end of his treatise, Galileo applies his new knowledge discovered through the experimental method to the improvement of artillery. So much for Bacon's general admonition. In the twentieth and the twenty-first centuries, laboratories worked for the benefit of life *and* the destruction of humanity, producing polio vaccine *and* thermonuclear weapons, aids for life *and* instruments of death.

In effect, Bacon rewrote the Bible. The Old Testament told how Adam and Eve were expelled from the Garden of Eden. Eve ate the forbidden fruit because she wanted to be like God. She knew that Adam not only loved her but also loved God, for her husband was absent from her when he walked in the Garden in the cool of the day with God. Eve desired to have all of Adam's love; she wanted no one to exist for Adam outside of her and believed that if she became like God, then she would be the sole recipient of all Adam's love.

Adam, too, desired to be like God. He wanted Eve to love him, nay, to worship him, as if he were the Lord of Creation. Adam wanted Eve to glorify him, to become the Lord God for her.

The Fall of Man was caused by misdirected love, not sex. Both Adam and Eve had an exalted self-love that excluded all other love. Each desired to be loved as if he or she were the center of all existence; both ultimately refused to love the other or God. But not to love the other and God is to separate oneself from the other and God. Thus, an exclusive self-love separated both Adam and Eve from each other and from God; however, nothing exists separately from God — to be separated from God is to fall into nothingness.

Adam and Eve, by choosing the love of self over the love of God, chose nothingness over the fullness of life with God in the Garden of Eden. God punished Adam and Eve for their disobedience by allowing them to receive what they chose — death. As a result, decay and mortality entered the human body and the material world, and the image of God in the human soul created good became tarnished.

The New Testament promised that what had been lost through the disobedience to God would be restored through faith in Jesus Christ. In Bacon's version of Genesis, the restoration of the lost Paradise was expected to come about not through faith but from the "great mass of inventions"[250] that would result from the new experimental science. The descendants of the first man and woman could ignore God and on their own return to Paradise to "subdue and overcome the necessities and miseries of humanity"[251] that resulted from the expulsion of Adam and Eve to the East of Eden, where women painfully suffered childbirth and the cursed ground brought forth thistles and thorns.

In the Cathedral of Science, the philosophy of materialism is preached since the toolbox of experimental science is limited to air pressure, chemical changes, electrical impulses in nerves, brain cell activity, and other measurable properties of matter. Said in terms of modern reductionism, "the universe, including all aspects of human life, is the result of the interactions of little bits of matter."[252] The amazing new inventions focused everyone's attention on the good life in this world, away from salvation, at the expense of intellectual and spiritual values, the usual definition of materialism.

Bacon transformed Judeo-Christianity into a secular, materialistic philosophy, and no doubt he would be aghast, for he vigorously opposed atheism: "I had rather believe all the fables in the Legend [a thirteen-century collection of saints' lives], and the Talmud, and the Alcoran [the Koran], than that this universal frame is without a mind."[253] He maintained that atheism reduces man to a mere animal: "They that deny a God destroy man's nobility; for certainly man is of kin to the beasts by his body; and, if he be not of kin to God by his spirit, he is a base and ignoble creature."[254]

The attempt to institute Paradise on Earth was a bold, hubristic project, entirely new to humankind, and unwittingly a vast social experiment that from the viewpoint of the twenty-first century was a catastrophic failure.

With the denial of God in the Second Fall, history had to be rewritten. Karl Marx constructed a coherent, rational view of history to replace the narrative of the unfolding relationship between God and Man in the Hebrew Bible. Marx eliminated God in favor of Man. A person could be free and independent only if God does not exist: "A man who lives by the grace of another regards himself as a dependent being. But I live completely by the grace of another if I owe him not only the maintenance of my life but also its *creation*, if he is the *source* of my life. My life necessarily is grounded outside itself if it is not my own creation."[255] In the future, humankind will be free because man creates man through economic production. "*The whole of what is called world history* is nothing more than the creation of man through human labor," Marx declared.[256]

Marx moved the Garden of Eden to the future. In *The Communist Manifesto*, Marx and Friedrich Engels described the future Paradise: "When, in the course of development, class distinctions have disappeared, and all production has been concentrated in the hands of a vast association of the whole nation, the public power will lose its political character. Political power, properly so-called, is merely the organized power of one class for oppressing another." In the grand vision of the *Manifesto*, history ends with the disappearance of oppressors and freedom for all: "In place of the old bourgeois society, with its classes and class antagonisms, we shall have an association in which the free development of each is the condition for the free development of all."

Instead of a Paradise, the social experiment in the Soviet Union to establish the Kingdom of Heaven on Earth produced a nightmare with the deaths of over 15,000,000 people; later, Chairman Mao with the Great Leap Forward would top the Soviets with the murder of 30,000,000 people. Maybe, Pol Pot should win the prize for total revolution. He and his Khmer Rouge restarted civilization in the Year Zero, the time when revolutionary ideals replaced all culture and traditions within a society. Slave labor, malnutrition, poor medical care, and executions killed a quarter of the population of Cambodia.[257]

In his monumental study, *The Gulag Archipelago*, Aleksandr Solzhenitsyn documented the torture and mass killings of the Soviet regime. At Solovki, guards poured cold water on the prisoners in subzero temperatures; in summer, naked prisoners were tied to trees to be devoured by mosquitoes. The prisoners at Zrossky Springs were so famished they ate a horse, dead for more than a week and covered with flies and maggots. At Utiny Goldfields, the prisoners ate half a barrel of lubricating grease. At Mylga, they ate Iceland moss, like the deer. In 1938, the NKVD (People's Commissariat of Internal Affairs) circulated the secret instruction to reduce the number of prisoners held in the camps by the use of disease, starvation, and murder.[258]

In the winter of 1931-1932, 100,000 prisoners died constructing the White Sea Canal. Dmitri Vitkovsky, a work supervisor, who saved many lives with the falsification of work records, describes the end of the workday:

"There were corpses left on the worksite. The snow powdered their faces. One of them was hunched over beneath an overturned wheelbarrow; he had hidden his hands in his sleeves and frozen to death in that position. Someone had frozen with his head bent down between his knees. Two were frozen back to back leaning against each other.... At night, the sledges went out and collected them. The drivers threw the corpses onto the sledges with a dull clonk."[259]

That is how 1,600,000 or more Soviet citizens died in the failed attempt to establish Paradise on Earth.[260]

The Nazi rewrite of history eliminated the Jews in favor of the Aryans and replaced the Garden of Eden with an Aryan Paradise on Earth.

To demonstrate that the Aryans were the chosen people, Adolf Hitler, Heinrich Himmler, and Adolf Eichmann murdered six million Jews and disposed of their bodies in mass graves and factory crematoria. The Nazi theologians argued, "How could the Jews be the chosen people when we use their ashes for tire traction on ice? Therefore, we are the chosen people. Q.E.D." Primo Levi, a survivor of Auschwitz, locates the foundation of Nazi doctrine as anti-Semitism. For the Nazis, he writes, "the Jews could not be 'the people elected by God,' since that's what the Germans were."[261]

To give scientific authority to his ideas on race, Hitler drew upon the social Darwinian principle that nature destroys the weak. For the Führer, all life was an eternal struggle, the world a jungle where the fittest survived and the strongest ruled — "the world of everlasting struggle, where one creature feeds on the other and where the death of the weaker implies the life of the stronger."[262] *Mein Kampf* is peppered with social Darwinian pronouncements: "In the end, the instinct of selfpreservation alone will triumph . . . Man has grown great through perpetual struggle. In perpetual peace, his greatness will decline. . . . The stronger must dominate and not mate with the weaker . . . Only the born weakling can look upon this as cruel . . ."[263]

To ensure racial purity, the Nazis, in 1933, introduced compulsory sterilization for congenital mental defects, schizophrenia, manic-depressive psychosis, and hereditary diseases such as epilepsy and chorea.

Konrad Lorenz, later to receive the Noble prize for co-founding the science of animal behavior with Nikko Tinbergen[264] and Karl von Frisch, gave, in 1940, the medical "argument" for racial purity: "There is a certain similarity between the measures which need to be taken when we draw a broad biological analogy between bodies and malignant tumors, on the one hand, and a nation and individuals within it who have become asocial, because of their defective constitution, on the other hand . . . The elimination of such elements is easier for the public health physician and less dangerous for the supraindividual organism, than such an operation by a surgeon would be for the individual organisms."[265]

Despite the Germanic wordy circumlocution, some physicians heeded Lorenz's ideas on racial purity, especially when he stated without ambiguity,

Figure 13.1. Officers of the Master Race murder two "subhumans." Public domain.

"It must be the duty of racial hygiene to be attentive to a more severe elimination of morally inferior beings than is the case today."[266]

Hitler declared, "Conscience a Jewish invention . . . a blemish like circumcision."[267] The revaluation of all values adopted from the philosopher Friedrich Nietzsche demanded that an ardent Nazi break free from constraints of conscience and the emotional barriers against cruelty. The central Nazi moral principle was inferior peoples are the raw material for the establishment of a glorious Paradise on Earth based on race. In such a future paradise, the Aryans will elevate human achievements to undreamed of heights. See illustration.

The Nazi attempt to establish an Aryan Paradise on Earth was a disaster for humankind.

Bacon imagined a glorious future for us once our ancestors made themselves the masters and possessors of nature; however, the ever-ascending arc of science and technology turned out to be not under human control. Physicists, neuroscientists, and computer and genetic engineers are the new sorcerer's apprentices, having summoned great forces that they now cannot either control or banish. Science and technology became the masters and possessors of us.

No one knows how molecular nanotechnology, genetic engineering, and artificial intelligence will transform human life, not the engineers at M.I.T., the geneticists at Stanford, or the computer scientists in Silicon Valley. Perhaps the ever-ascending arc of science and technology is headed to super-intelligence, maybe to a thermonuclear war that annihilates humankind, or possibly to a severe climate change that destroys *Homo sapiens* and most other creatures.

Even if the terrible wars and political crimes against humanity of the twentieth century belong to a darker age, the West still faces the brick wall of personal discontent and social disintegration — youth suicide, drug addiction, and violent crime plague every industrial society. The educated and well-off surrounded by affluence are prone to loneliness, depression, and malaise. Bacon's secular vision to create through science and technology a Paradise on Earth ended in a series of social and political catastrophes. Hiroshima killed the comforting narrative that the progress of science and technology leads to universal happiness; the Gulag destroyed the utopian story of humans instituting a perfect political order; the Holocaust annihilated Nietzsche's myth that the ascendance of the Overman, the perfect race, will raise humankind to a new level of existence. The grand storytelling of Post-Christianity is over.

14 The Great Givens of Human Life

I don't know about you, but I wish I had been born with the user's manual for the human being, or at least received basic instruction in school about the fundamental equipment every human being possesses. In my youth, I was incredibly ignorant about my senses, emotions, memory, imagination, intellect, and will. For me, it was like waking up alone and finding myself on board the space shuttle, and, then, through trial and error trying to figure out how to fly the machine, without much success. After a series of personal disasters that are irrelevant to report here, I realized that all life entails givens. Biological life on earth rests on sunlight, the source of high energy with low entropy. Plants live on light, and all animals, directly or indirectly, live on plants. Certain elements of human life are also givens that are outside the realm of personal acceptance or rejection; they just are.

Any person can briefly reflect upon his or her life and discover the great givens. For example, I was born into a family, and, of course, families do not exist by themselves. My family did not speak a private language; when family members spoke to me, I was connected to a larger community that shared a common life. Hence, one given of human life is that we are always connected to other persons; we are individuals *and* members of a group. No human being is a solitary creature. Everyone necessarily belongs to a family and a larger community, speaks a language, and shares opinions and beliefs, joys and sorrows.

Although my capacity to speak a language was given to me by nature, that I spoke English and Romanian was given to me by culture. Therefore, every human person necessarily lives in two worlds, the natural and the cultural. Even the most confirmed city-dweller, who thinks milk comes from cartons and who believes that nature gives us nothing for free, must at times feel the sun upon her face and experience the wonder of rebirth in spring.

Experience taught me that I live in a world of objects, yet not everything that I experienced could be touched. My emotions and thoughts were intangible. Each person, thus, has two lives, a life in the physical world and an interior life.

That I lived in time was obvious to me as a child: I went to bed at night and got up in the morning. I saw that the world around me always changed. When I was two, my favorite toy was a horse mounted on a wheeled platform. A leather strap was tied to the horse's neck, and when I pulled the strap, the horse followed me. I spent hours playing with my horse. Then, one day, a wheel fell off the horse, and when no one could fix it, I cried for hours. I learned no material thing is permanent.

Only in high school, when I saw the demonstration that the prime numbers are infinite, did I experience timelessness. Then, I saw that the goal of science is to discover the unchanging in a world of flux. Science seeks the timeless. In some mysterious way, every person lives in time and yet transcends time.

Human life, then, is made up of great pairs of opposites. The elements of a complementary pair are yoked together; neither exists without the other. Life, itself, arises from the dynamic tension between polar opposites. Generation results from male/female; sensory life from light/dark, sound/silence, hot/cold, sweet/sour, pleasure/pain; the interior life from love/hate, joy/sorrow, true/false, free/determined, changing/unchanging; human action from good/bad, individual/group, man-made/natural.

These fundamental polarities are the great givens of human life and are acknowledged by all cultures. What defines a culture is how it develops these polarities in each of its members. Traditional Hopi Indian culture, for instance, is matriarchal and emphasizes the group over the individual, nature over the man-made, the unchanging over the changing. In contemporary America, the extreme emphasis on the individual determines how all the other polarities are developed. Isolated, autonomous individuals ignore nature, compete with others, see freedom as a supreme good, disregard both the remote past and the distant future, and since all individuals are deemed equal, the differences between the male and the female are blurred

We Are Born Radically Incomplete

Life is an activity directed toward an end; however, plant, animal, and human life are not guided by the same principles. Although plant and animal physiology differ, all multi-cellular organisms grow from "seeds," metabolize nutrients, and reproduce. Besides this vegetative life, the higher animals have an interior life; they perceive an outside world, feel pleasure and pain, and act out of instinct and emotion.

We thus have much in common with the animals. To see how much we differ from them, imagine six animals queued up for scrutiny, much like in a detective movie. In our animal lineup, we have an anteater, a zebra, a rhino, an astronaut, a stockbroker, and a sumo wrestler. Nature furnishes the anteater, the zebra, and the rhino with a fixed occupation, a fitting dress, and an appropriate emotional profile, respectively. The long, tapered snout of the anteater is its tool for carrying out its occupation, eating ants and grubs. No two zebras have identical stripes. A baby zebra identifies its mother by her stripes. Lions and other animals that prey on zebras are confused by the flashing stripes displayed by a running zebra herd. The dress of the zebra fits its way of life. An adult male rhino, in mating season, marks out territory with its urine. A rhino occupies the center of its territory and aggressively chases away any male rhino that challenges it. A male rhino, however, must leave its territory for water and then out of necessity it crosses the territories of other adult males. When a rhino intrudes into another rhino's territory for water, it becomes submissive. The farther a rhino strays from the center of its territory, the more submissive it becomes. The aggression of the male rhino is regulated by nature.[268]

Except for *Homo sapiens*, all the animals in our detective lineup have a complete life given to them by nature, although some animals do possess a limited culture. For instance, unlike a mallard or a robin, a young jackdaw, a medium-sized member of the crow family, does not recognize a cat as an enemy by instinct. The jackdaw learns that a cat is an enemy, not through its own experience of a cat attacking it, or even by witnessing a cat attack another jackdaw, but by "actual tradition, by the handing-down of individual experience from one generation to the next!"[269] But the culture of any animal is so circumscribed that virtually all its behavior is rigidly determined by nature, as animal trainers know.

Psychologists Keller and Marian Breland, the proprietors of Animal Behavior Enterprises, have had more than fourteen years' experience in training animals for various commercial purposes.[270] Whenever the Brelands attempted to train an animal to go against its instinctive behavior, they met with persistent failure. Chickens trained to deliver to a spectator a plastic capsule containing a toy would, after a few successful performances, began to stab at the capsules and pound them up and down on the floor of the cage. Pigs trained to deposit large wooden coins in a piggy bank for immediate food rewards would do well for a few weeks but then began dropping the coins repeatedly, rooting them, tossing them into the air, and rooting them again indefinitely. Chickens that hammer capsules are obviously exhibiting instinctive behavior having to do with the breaking open of seed pods or with the killing of insects and grubs. The rooting behavior of pigs is part of their food-getting behavior. Animals can deviate little from the course set for them by nature; consequently, successful animal training builds upon instinct.

In our strange collection of animals, those in our detective lineup, *Homo sapiens* is the strangest of all. Nature gives human beings no specific way of life — no fixed occupation, no fitting dress, and no appropriate emotional profile. The life of an anteater, a zebra, or a rhino is infinitely easier than the life of a human being, although infinitely smaller. We humans must struggle to find an excellent way of living, if it exists, but we have a vastly richer interior life — nothing great without a curse.[271]

What nature gives the other animals, we acquire through language, the tool that defines *Homo sapiens*. But we are born only with the potential for language that must be actualized through culture; we must be completed by others.

We Are Social by Nature

Without others, we could not acquire language, and without language, our mental life would not be much higher than that of a chimpanzee or a bonobo. Helen Keller gives us a glimpse of how the world is experienced in the absence of language. When she was 19 months old, an acute disease, possibly scarlet fever or meningitis, left her blind and deaf. She soon "felt the need of some communication with others and began to make crude signs. A shake of the head meant 'No' and a nod, 'Yes,' a pull meant 'Come' and a push 'Go.' Was it bread [she] wanted? Then [she] would imitate the acts of cutting the slices and buttering them."[272]

Without language, Helen's interior life was limited to sense perception, motor skills, tactile memory, and associations. She exercised neither will nor intellect and was "carried along to objects and acts by a certain blind natural impetus." She felt anger, desire, and satisfaction; however, she never "loved or cared for anything." She describes her inner life then as "a blank without past, present, or future, without hope or anticipation, without wonder or joy."[273]

Ms. Keller reported that the sign language she learned from Anne Sullivan "made me conscious of love, joy, and all the emotions. I was eager to know, then to understand, afterward to reflect on what I knew and understood, and the blind impetus, which had before driven me hither and thither at the dictates of my sensations, vanished forever."[274] She no longer lived an animal life; language freed her to be human.

Without language, without others to learn language from, the mental capacities that Ms. Keller, you, and I were born with would have not developed, and our lives would not have been much higher than that of an animal. Through language, humans bring out the full potentiality hidden in matter, advance the building of bird nests and beaver dams to architecture and engineering, the gathering of nuts to farming, squawks and barks to music, sexual reproduction to love and compassion, and limited animal perception to the intellectual jewels of modern Western culture, Newtonian physics, Maxwell's electrodynamics, special and general relativity, quantum physics, and the biology of the physical basis of life, including the genetic code.

Even the newborn infant reveals the social nature of *Homo sapiens*. In 1961, ethologist Robert Fantz developed a reliable technique for measuring the visual preferences of babies. Presenting a reclining infant with two visual stimuli, he measured the amount of time each object was reflected in the infant's pupils. In this way, Fantz was able to infer the baby's preference for one object over another. It is now known that newborn vision is at least 20/150, an acuity not exceeded by many adults: "By demonstrating the existence of form perception in very young infants [Fantz] . . . disproved the widely held belief that they are anatomically incapable of seeing anything but indistinct blobs of light and dark."[275]

Fantz and many subsequent experimenters found clear evidence that babies, even those less than twenty-four hours old, prefer to gaze at a human face more than any other object, whatever its color, shape, or pattern. Other investigators found that "the human voice, especially the higher-pitched female voice, is the most preferred auditory stimulus in young infants."[276] These preferences are clearly not learned: In one study, the youngest babies were ten minutes old.

In other experiments, Fantz showed that without learning or experience, a newly hatched chick prefers to peck at three-dimensional, round, small objects. Nature directs the chick to look for grain. Similarly, as soon as the human infant emerges from the womb, it looks for a human face and listens for a soprano voice. Nature directs the infant to seek its mother. The very first experience in any person's life is connecting himself or herself to another person.

We Are Rational Beings

Because we humans are unfinished, we are not enslaved to anatomy the way animals are. The natural tools, weapons, and armor of animals serve only one specific task and cannot be put aside or changed for others, severely restricting the life of an animal to limited activities. A mole's short, chunky paw is an outstanding digging tool, but it cannot hold anything. An eagle's talons are perfect for clutching small animals but are useless for digging. The human hand can perform all the tasks achieved by the restricted tools of animals: It can dig with a hand shovel, stab with a sword, cut down a tree with a saw, and perform thousands of other activities without being restricted to any single one of them.

Unlike the instincts and organs of other animals suitable only for specific tasks that lock them into one way of life, the human mind and the human hands are general tools. The human hand is tailored to the human mind. Planting a garden, making a canoe, and painting a picture are rational activities; in each case, to achieve a desired end, the hands move under the direction of the mind. Thus, human activity is rational.

The moral life is a particular type of rational activity. For now, let us consider only eating. By instinct, animals avoid eating harmful things. Ethologist James Gould writes, "Many animals from blue jays to garden slugs come programmed to wait a species-specific length of time after eating a new food to see if they become ill. If they do — even if the sickness arose from a completely independent cause — they will never eat the food again. Even more curious, each species is programmed to identify the forbidden food in the future by its own set of cues. For example, rats will remember the suspect food's *odor*, while quails recall its *color*."[277] Instinct also directs animals in how much to eat. A biology professor friend of mine keeps a boa constrictor in his lab. After the boa eats a rabbit, it will not touch another live rabbit until it digests the one it has eaten. By nature, the boa is a temperate eater.

I remember — and I know my son would like me to forget — that once or twice when he was young, he threw up after having gorged himself on Halloween chocolate candy. In that way, he learned that a human being must use reason to direct his desires and to acquire temperate habits.

By Nature, We Desire to Know

Confucius said, "I am a person who forgets to eat when in the pursuit of knowledge, forgets all worries when he is in the enjoyment of it, and is not aware that old age is coming on."[278] Even infants show an irrepressible passion for investigating the world. They crawl about, raiding cupboards, experimenting with all manner of objects, and tasting everything. Such behavior is clearly natural, not taught.

As soon as a child learns to speak, an unending barrage of questions begins, as every parent knows. Children are experts at wondering, for they see the world with new eyes. Some of us preserve the spontaneous wonder we had as children. Astronomer Sir Arthur Eddington points out how childlike wonder animates the scientist: The child who sings *Twinkle, twinkle little star, How I wonder what you are,* "is wondering how big it is and how far away, what keeps it from dropping down, whether it is made of gold, whether it is lit by electricity. . . . One question leads to another, and in the recondite treatises of physics we are still asking, and now and again answering, the unceasing flow of questions."[279]

The actual life we live first as children and then as young adults — if we did not have our questioning turned off by parents and teachers *and* if we do not become corrupted by the desire for power, wealth, and material comfort — shows that one purpose of human existence is to reveal, or to uncover, or to come into contact with more and more truth. Just as the sunflower cannot help but turn from the dark toward the light, the human being cannot help but shun falsehood and seek the truth.

Thus, we do not determine the most important thing about who we are: our natural end in life. Trying to fulfill our rational nature through loving and seeking truth, we become more and more. A desire that would be thwarted if our rational nature did not mean that we possess virtually an unlimited intellectual freedom. If our understanding were determined by instinct, brain physiology, and culture, then we could never separate the true from the false and knowing would be impossible. For science to be possible, the scientist must have the freedom to choose, and what holds for the scientist also applies to the layperson; *every human has the capacity to make free choices.*

The Senses Need Training

Knowing begins with the senses; unlike the animals, however, human senses require training. The untutored tongue cannot distinguish a St. Emilion from a St. Julien, though a wine enthusiast not only can recognize the region the wine came from but the chateau and the vintage year.

The education of the whole person must include training in the use of the senses. Chief Standing Bear says that Lakota children, "were taught to use their organs of smell, to look when apparently there was nothing to see, and to listen intently when all seemingly was quiet. A child who cannot sit still is a half-developed child."[280]

In the one-room schools of America's past, observation lessons were often given. John T. Prince, in his *Courses and Methods: A Handbook for Teachers*, written in 1892, explains that the "aim of these lessons is not so much to teach facts as it is to cultivate the pupils' powers of observation, and to awaken an interest in, and a love for, the things of nature that lie directly about them." Prince suggests that teachers tell nothing to their students that they can discover for themselves by their own powers of observation. He maintains that for students, "Learning one fact by their own unaided powers is better than memorizing a hundred facts which have been given to them."[281]

Age is no barrier to training the senses. An adult can see the parts of a flower — the sepals, petals, stamens, and pistils — or observe that the head of the common ant is triangular in shape, or learn how to use the pointer stars of the Big Dipper to locate the North Star.

While reflecting upon Lakota life, Standing Bear discovered the universal truth that "half-dormant senses mean half living."[282] When a person's senses are alive and alert to the world around him or her, life is full and interesting. Of all the natural creatures, only human beings can perceive the fullness of nature. Who does not wonder how birds fly, why the trees turn color in the fall, how ants find their way back home, or why heavy objects fall? If a person is alive, then everything in nature evokes wonder.

The Joy of Knowing

Every naturalist enjoys the pleasure of exercising his or her powers of observation. Listen to Charles Darwin's journal entry describing his first day in a tropical jungle: "The day has passed delightfully. Delight itself, however, is a weak term to express the feelings of a naturalist who for the first time has wandered by himself in a Brazilian forest. The elegance of the grasses, the novelty of the parasitical plants, the beauty of the flowers, the glossy green of the foliage, but above all the general luxuriance of the vegetation, filled me with admiration. A most paradoxical mixture of sound and silence pervades the shady parts of the wood. The noise from the insects is so loud that it may be heard even in a vessel anchored several hundred yards from the shore; yet within the recesses of the forest a universal silence appears to reign. To a person fond of natural history such a day as this brings with it a deeper pleasure than he can ever hope to experience again."[283]

The beauty of nature connects a Lakota Indian Chief, a great English naturalist, and a pupil in a one-room school house on the American prairie. Fully developed senses allow us to receive the gifts of nature: beauty, wonder, mystery, and places to meditate — the means to discover that we belong in this world as much as the wild sunflowers and the soaring hawks.

The Good Mind

The study of music, language, literature, mathematics, and science develops our capacity to define, analyze, and draw conclusions. For these studies to bear fruit, we must acquire more than knowledge, techniques, and general rules. We must be trained to think well, and this is possible because we are unfinished by nature and thus must perfect ourselves. A good mind is open, thinks concretely, and seeks interconnections.

Openness

The open mind willingly accepts truth from any source. Mozart, a model of open-mindedness in music, writes, "People make a mistake who think that my art has come so easily to me. Nobody has devoted so much time and thought to composition as I. There is not a famous master whose music I have not studied over and over."[284] Mozart's reaction to Bach's music reveals his childlike openness. When one choirmaster and his forty-pupil choir performed for Mozart an eight-part motet by Bach, the music had scarcely began, "before Mozart started with an exclamation, and then was absorbed in attention. At the conclusion he expressed his delight, and said, 'That now is something from which a man may learn.'"[285]

Close-mindedness often arises from laziness or disdain. If a person rests contentedly with his opinions, how can he learn from others? Similarly, the stubborn mind desires to refute any one on any topic, a major obstacle to learning. An open mind, in contrast, recognizes that all of us are profoundly ignorant, and thus easily admits that it does not know much — nobody does!

If we admit our ignorance to ourselves, we will see that our opinions carry little weight and thus need to be examined, especially culturally-given opinions that most of us take as obviously true. If we are willing to confess our ignorance in public, we can then ask questions openly and engage in genuine dialogue. If we are constantly aware of our ignorance, then we will always have the freshness and innocence of a beginner, who is astonished again and again by the new wonders he or she encounters. We will never forget that all learning begins with wonder and amazement, and that profound truth appears strange to cultural opinion.

Concreteness

In thinking, concreteness yields clarity. One of the maladies of modern life is the substitution of a fuzzy verbal world for actual, concrete experience. Psychologist Abraham Maslow observes that genuinely creative thinkers "live far more in the real world of nature than in the verbalized world of concepts, abstractions, expectations, beliefs, and stereotypes that most people confuse with the real world."[286]

Physicist Enrico Fermi was noted for his quick and clear thinking. One reason for his mental agility and clarity of thought was he had "a whole arsenal of mental pictures, illustrations, as it were of important laws or effects."[287] He would not simply keep Newton's third law in his head, for example. (To every action there is an equal and opposite reaction.) He would discover and etch in his memory a paradigmatic example of the law, such as a man jumping from a boat to a dock where clearly the boat must move away from the dock with a momentum equal to the man moving toward it.

If we cannot give a simple, obvious example of something, we probably do not know what we are talking about. Using this principle to assess ideas, theories, books, lectures, either our own or those of others, enables us to cut through extraneous matters to the essentials in rapid fashion.

Interconnections

Without seeking interconnections as we learn, what fragile knowledge we gain can be easily lost. We, thus, should form the habit of connecting what we are learning to what we already know. For example, when we hear Stephen Daedalus, the protagonist of James Joyce's A Portrait of the Artist as a Young Man, argue that the three universal qualities of beauty in the arts are wholeness, harmony, and radiance, his translation of Aquinas' integritas, consonantia, and claritas, we should seek to see if this trio describes beauty in the sciences. Einstein does give these three elements: "A theory is the more impressive the greater the simplicity of its premises is, the more different kinds of things it relates, and the more extended its area of applicability."[288] As a further confirmation that beauty is a common ground that unites the arts and the sciences, physicist and novelist C. P. Snow writes, "The literature of scientific discovery is full of aesthetic joy. The very best communication of it that I know comes in G. H. Hardy's book, A Mathematician's Apology. Graham Greene once said he thought that, along with Henry James's prefaces, this was the best account of the artistic experience ever written."[289]

The Curse of Social Living

My mantra "nothing great without a curse" adopted from Sophocles' *Antigone*, applies to the social nature of *Homo sapiens*, too, not just to the human condition of nature not giving us a specific way of life. Individually, each one of us is extraordinarily weak and could not survive on our own. Even a recluse who retires to a remote region of Alaska to live alone brings with him knowledge and skills acquired from prior group living. In our highly technological society, no person knows how to produce everything that he or she consumes or uses in a single day. What person knows how to grow broccoli, make eyeglasses, weave cloth, generate electricity, and fabricate a microchip? The community of humans supplies all our needs. The farmer is given the fruits of ten thousand years of experimentation with the growing of crops; the poet, a language and the poetry of Homer, Dante, and Shakespeare; the physicist, the understanding of Newton, Einstein, and Bohr. No farmer, no poet, or no physicist could ever pay for all the gifts he or she receives gratuitously. Each one of us can humbly accept what is freely given, preserve and add to it if possible, and then pass it on to others. When we understand ourselves as parts of a whole and recognize that our lives are possible only because of the group, we see that such destructive emotions as anger, envy, and self-pity are contrary to our social nature, and we willingly work for a community life that promotes peace, cooperation, and generosity.

The curse of social living is that every society implants ideas and instills habits of feeling and thinking that limit its members to a particular perspective, one that as a general rule is contrary to human nature and destructive to neighboring societies. The paradox is that social living greatly extends our capabilities and yet limits us. Capitalism tells us that we are economic beings, consumer-workers. Nationalism tells us that our ultimate destiny is the fate of our Nation-State. Democracy tells us we are autonomous, isolated individuals.

In my youth, I believed I was an island unto myself and that I had freely chosen my own way of life with no regard to what others thought of my odd, eccentric behavior. Then, I read Alexis de Tocqueville's *Democracy in America* and discovered that the individualism of the New World inculcated Cartesian thinking in me; as a result, like a trained parrot, I vociferously and probably obnoxiously tried to convince my friends in philosophy and literature that they were fundamentally deluded, for the universe, including their own misguided thinking, could be explained by tiny bits of matter. Tocqueville, not my friends, persuaded me that I was an idiot, intellectually unaware how modern culture shaped my interior life.

Culture gives every person the tools for human living as well as a core self. I had wholeheartedly believed the democratic myth that I was the King of the Castle, that I had chosen the fundamental aspects of how I lived, and that I was totally free, unrestrained by a non-existent human nature. I probably would have gone on forever believing this nonsense — no, living this nonsense — if it were not that on rare occasions life awakens us more than ideas in books.

As a post-doc in theoretical physics at Los Alamos National Laboratory, I heard Samuel Glasstone deliver twenty hours of classified lectures on the history of nuclear weapons. Hiroshima and the development of the hydrogen bomb — the sheer destructiveness of nuclear weapons and their insanity as military strategy — changed everything for me. For the first time in my life, I saw that my life was not mine alone, that what I choose to do would affect others, and that I was inextricably bound to others, even to all humanity. In the course of my crazy, zigzag life, I arrived at the place where John Donne had been four hundred years before me. "No man is an island, entire of itself; every man is a piece of the continent, a part of the main. . . . any man's death diminishes me, because I am involved in mankind, and therefore never send to know for whom the bell tolls; it tolls for thee."[290] And, I heard the bell toll for me, and left Los Alamos, never to return.

15 *Homo sapiens*: A Wondrous Creature

Several years ago, Fortuna, an old Indian woman, invited my wife and me to a dance at the San Ildefonso Pueblo, about twenty miles from Santa Fe, New Mexico. At the beginning of the ancient ritual, the Hunters with sprays of evergreen at wrist and knee stood in a line opposite the Hunted, stripped to their waists and painted with symbols, and wearing headdresses of green twigs and horn. Fortuna explained that the Hunters renounced enmity toward all animals and begged the Hunted to sacrifice their lives so human beings could continue living. Toward the end of the dance, I recalled that Thomas Aquinas argued that the powers of cogitation and memory are not absent in animals, but they are more perfect in humans.[291]

Animals Are Not Mere Machines

I knew that unlike Aquinas and the Pueblo Indians, "most biologists and psychologists tend, explicitly or implicitly, to treat most of the world's animals as mechanisms, complex mechanisms to be sure, but unthinking robots nonetheless."[292] Descartes proposed that animals were nothing but intricate machines and saw no need to invoke conscious awareness to explain their behavior: "Since art is the imitator of nature, and since man is capable of fabricating various automata in which there is motion *without any cogitation*, it seems reasonable that nature should produce her own automata, far more perfect in their workmanship, to wit all the brutes."[293]

Descartes developed a strictly mechanistic theory of animal behavior, postulating that invisible but corporeal particles are excited by the senses and course through tiny pores into the brain, where they flow through nerves to instigate muscle contraction, causing movements of an animal. The entire process is mechanical without the intervention of the awareness of the animal.[294] He compared the animal body to a fountain statue which, powered by hydraulic mechanisms, give the semblance of self-initiated movement.[295] Descartes furthered argued that there is no more need for a soul in an animal to explain its movements than for "a soul in a clock which makes it tell the time."[296] All takes place according to the disposition and arrangement of the animal's parts; only the human soul thinks, feels, and perceives; no other animal has an interior life.[297]

According to animal-rights activist, Gary Francione, "Descartes and his followers performed experiments in which they nailed animals by their paws onto boards and cut them open to reveal their beating hearts. They burned, scalded, and mutilated animals in every conceivable manner. When the animals reacted as though they were suffering pain, Descartes dismissed the reaction as no different from the sound of a machine that was functioning improperly. A crying dog, Descartes maintained, is no different from a whining gear that needs oil."[298]

In the nineteenth century, Descartes' theory became biology's investigative program for animal behavior. Biologist Thomas H. Huxley wrote an essay, in 1874, defending the Cartesian hypothesis: "Brute animals are mere machines or automata, devoid not only of reason, but of any kind of consciousness. What proof is there that brutes are other than a superior race of marionettes, which eat without pleasure, cry without pain, desire nothing, know nothing, and only simulate intelligence as a bee simulates a mathematician?"[299] (See Figure 15.1.)

Figure 15.1. Every morning a coyote snoops around my garden. In a Cartesian cosmos, Wile E. Coyote is a machine that eats without pleasure, cries without pain, and desires nothing. (R. H. Barrett, U.S. Fish & Wildlife Service, public domain, Wikimedia Commons.)

Huxley maintained that even if awareness exists in animals, it would arise only as a side effect of the mechanism of their bodies and would be incapable of causing anything in the animal's behavior: "The consciousness of brutes would appear to be related to the mechanism of their body simply as a collateral product of its working, and to be as completely without any power of modifying that working as the steam-whistle which accompanies the work of a locomotive engine is without influence upon its machinery. Their volition, if they have any, is an emotion indicative of physical changes, not a cause of such changes."[300]

Today, the Cartesian program has been extended to include humans. After the mapping of the human genome, the new rage in evolutionary psychology is to explain all human behavior in terms of genes and brain function, not just physical traits such as left- or right-handedness and tongue curling, folding, and rolling, but moral virtues such as generosity, courage, and temperance. Genomania purports to explain all human behavior by genetic variation. According to the new phrenologists, Evel Knievel attempted unsuccessfully to vault his Harley-Davidson XR-750 over thirteen parked Pepsi trucks because he possessed a gene that increased risk-taking.

Assistant professors seeking tenure, researchers pursuing a Nobel Prize, and evolutionary psychologists hoping to be the next Darwin are kicking the bushes in Cambridge, Massachusetts, Ann Arbor, Michigan, and Berkeley, California, in search of the genes that determine temperament, emotional responses, levels of aggression, and, of course, our most revered moral choices. For most neuroscientists and evolutionary psychologists, human consciousness is like "the steam-whistle which accompanies the work of a locomotive engine, [but] is without influence upon its machinery."[301]

Alexis de Tocqueville, in 1840, made the shrewd observation that once materialists "think they have sufficiently established that they are no better than brutes, they seem as proud as if they had proved that they were gods."[302] Undoubtedly, in the twenty-first century, Tocqueville would have written "no better than brutes or a collection of genes."

To rid biology of what he perceived as the two principal errors introduced by Descartes — *Homo sapiens* is an exceptional animal and the other animals have no more of an interior life than a clock does — entomologist Edward Wilson introduced, in 1975, "the new discipline of sociobiology, defined as the systematic study of the biological basis of all forms of social behavior, in all kinds of organisms, including man."[303] For complicated reasons that may have more to do with intellectual marketing than with science, sociobiology became evolutionary psychology. The attempts of biologists to shake off the philosophical heritage of Descartes have borne unexpected fruit.

When I was an undergraduate studying physics and mathematics at the University of Michigan, I took Anthropology 101 from Leslie White. Somewhere in my class notes, lost years ago, is the sentence "The most satisfactory definition of man from the scientific point of view is probably Man the Tool-maker,"[304] a definition proved wrong by Jane Goodall fifteen years before I heard it.

(An aside: Much of what we absorb as the truth as naïve students is material understood from a narrow perspective and pounded into our heads through a system of examinations and rewards. In this case, an anthropologist defines *Homo sapiens* physically as the tool-making animal. But in every culture, objects are sculpted, songs are sung, and dances are performed, all artifacts. Why choose the making of stone axes and bone scrapers to define humankind?)

Tool-making Chimpanzees

As a twenty-six-year-old woman with no research experience, Goodall ignored the prevailing opinion that no human could ever grow close to wild chimpanzees and went to Gombe Stream National Park, Tanzania, in 1960, to study the species that turned out to be closest to *Homo sapiens*. Shortly before her year's research grant was up, she observed a male chimpanzee, one she called David Graybeard, insert stalks of grass into termite holes and then remove them from the hole covered with termites that he relished eating. On several occasions, David Graybeard "picked small leafy twigs and prepared them for use by stripping off the leaves. This was the first recorded example of a wild animal not merely *using* an object as a tool, but actually modifying an object and thus showing the crude beginnings of tool-*making*."[305]

Informed by a telegram of her discovery, Goodall's mentor, Louis Leakey, was "wildly enthusiastic" and arranged for the National Geographic Society to grant funds for another year's research.[306] Leakey said, "We must now redefine man, redefine tool, or accept chimpanzees as human!"[307]

Goodall's observations at Gombe challenged more than the definition of *Homo sapiens*. Biologists knew that certain birds use twigs, sticks, or cactus spines to dislodge insects and grubs from trees, the most famous being the woodpecker finch of the Galapagos Islands. Such tool-using behavior was assumed to be instinctual, a view later confirmed by a 2001 study that demonstrated that the use of twigs and cactus spines by the woodpecker finch is not acquired through social learning, since juveniles were able to use tools without having any contact with adults.[308] The researchers "found no evidence that woodpecker finches, in contrast to chimpanzees, learn tool-use socially."[309] Goodall had observed that the "actual tool-using patterns practiced by the Gombe Stream chimpanzees are learned by the infants from their elders."[310] If we define culture as a "way of doing" that is transmitted to young by imitating their parents and others, not by genetics, then chimpanzees and possibly other nonhuman primates have a culture.

Humans and chimpanzees obviously have in common all the defining characteristics of the genus mammal, but surprisingly our distant cousins and we share approximately 99 percent of our DNA[311] and the capacity to make tools. In Goodall's phrase, chimpanzees are "in the shadow of man," or in Aquinas' terminology, humans possess tool-making in a more perfect way. That we humans can bring out the full potentiality hidden in matter, advance the building of bird nests and beaver dams to architecture and engineering, and the gathering of nuts to farming needs no elaboration.

A more extended definition of culture includes the habits of feeling as well as doing that are passed on from one generation to the next, in contrast to instincts, which are genetically determined. Hence, to see if nonhuman primates have culture in a fuller way, we must examine the emotional life of animals.

Charles Darwin published, in 1872, *The Expression of the Emotions in Man and Animals*. His goal in this comprehensive book is to disprove that feelings are not unique to the human being by demonstrating that animals and man express rage, joy, fear, affection, suffering, and other emotions in the same way. To take just the case of anger. "When a dog is on the point of springing on his antagonist, he utters a savage growl; the ears are pressed closely backwards, and the upper lip is retracted out of the way of his teeth, especially of his canines."[312] Humans, too, in anger display their canine teeth, and when sneering give a "thoroughly canine snarl."[313] Angered chimpanzees also exhibit bared canine teeth. Unlike dogs and chimpanzees, humans can modify their instinctual expression of anger: "While the innate expression of anger involves baring of the teeth as in preparation for biting, many people clinch their teeth and compress their lips as though to soften or disguise the expression. People of different social backgrounds and different cultures may learn quite different facial movements for modifying innate expressions."[314]

Since Darwin's day, primatologists learned that unlike monkeys and virtually all nonhuman primates, chimpanzees console others in distress and seek consolation themselves when frightened or stressed. Goodall noted as a general rule, "When a chimpanzee is suddenly frightened, he frequently reaches to touch or embrace a chimpanzee nearby . . . Both chimpanzees and humans seem reassured in stressful situations by physical contact with another individual."[315]

Michael Seres, a research technician at the Yerkes Field Station in Lawrenceville, Georgia, repeatedly witnessed that "once the dust has settled after a fight [between two chimpanzees], combatants are often approached by uninvolved bystanders. Typically, the bystanders hug and touch them, pat them on the back, or groom them for a while. These contacts are aimed at precisely those individuals expected to be most upset by the preceding event."[316] Frans de Waal, a primatologist, amassed examples of the empathic capacity of the chimpanzee and its closest relative, the bonobo: Juveniles interrupted their rambunctious play each time they got close to a terminally sick companion; an old male led a blind female around by the hand; an adult daughter brought fruit down from a tree to her aging mother, who was too old to climb.[317]

Unlike the limited empathy of chimpanzees, human acts of kindness extend beyond the troop and the species and are incredibly numerous, ranging from animal lovers rescuing abused cats and dogs to an entire village in the French Alps saving thousands of Jews from extermination in Nazi death camps.[318]

Although chimpanzees mostly eat fruit, they love the taste of meat. Their principal source of meat is the red colobus monkey. In the 1990s, the chimps in an area called Ngogo in Kilbe National Park, Uganda, were killing up to half the red colobuses every year. As a result, their population fell by 89 percent between 1975 and 2007, as documented by two anthropologists, David Watts and John Mitani.[319] "I don't think chimpanzees are capable of thinking about a long-term future," says Watts. "They're just responding to what they encounter and what they see."[320] Unlike the chimpanzees, we can think about the long-term future of the earth and its species, but usually we do not — witness the extinction of the passenger pigeons and the coming disasters of climate change.

Violence is constantly present in chimpanzee life. The leading cause of death both at zoos and in the wild is infanticide.[321] When a female with an infant attempts to joins a new troop, she is allowed in, but her baby is killed by the adult males of her new community. Goodall observed the dark side of chimpanzee life. In a letter from the Gombe Stream National Park to her family back home in England, she described a "barbarous murder" committed by two female chimpanzees, mother and daughter. The two seized a three-week-old chimpanzee from his mother's arms and "deliberately killed [him], biting into his forehead. Then, the cannibalistic family fed on his remains for five hours."[322]

Waal reports that one troop of chimpanzees in the Gombe Stream National Park split into two communities, eventually becoming two separate troops. "These chimpanzees had played and groomed together, reconciled after squabbles, shared meat, and lived in harmony. But the factions began to fight nonetheless. Shocked researchers watched as former friends now drank each other's blood. Not even the oldest community members were left alone. An extremely frail-looking male, Goliath, was pummeled for twenty minutes and dragged about. Any association with the enemy was ground for attack."[323]

Sadly, humans have perfected violence. In the twentieth century, Nation-States, large and small, waged war among themselves, some against their own citizens. The power of the new gods enhanced by modern science and technology produced a century of political murder. A partial, conservative catalog of the horrors is mind-boggling, unbelievable, but undeniable. Deaths: World War I (military only): 9,700,000; Russian Revolution and Civil War: 9,000,000; forced collectivization: 4,000,000 Ukrainian peasants; Russian gulags: 1,700,000 political prisoners; Spanish Civil War: 1,200,000; World War II (military and civilian): 51,000,000; Nazi camps: 6,000,000 Jews and 6,000,000 Slavs, Gypsies, and political prisoners; Japanese Rape of Nanking: 300,000 Chinese; Allied bombing of Hamburg, Berlin, Cologne, and Dresden: 500,000 German civilians; Hiroshima and Nagasaki: 210,000 Japanese civilians; Vietnam War (military and civilian): 5,000,000.[324]

I used to think that except for *Homo sapiens* nature gave every species a complete way of life. My paradigm example was the rhinoceros. As we saw in the previous chapter, by nature, a rhinoceros "knows" what to eat, how to live in a herd, and when to breed. Nature even regulates a rhino's emotions to suit its needs.

Not All Animals Are Rigidly Determined by Nature

No wonder I thought that every animal has a complete life rigidly determined by nature, but the baboons proved me wrong. Robert Sapolsky, a primatologist and neuroscientist, spent many summers in the Serengeti ecosystem in East Africa. He admits that baboons are "often violent and abusive, so that the weak suffer at the canines of the strong."[325] One summer, the baboon troop adjacent to his study group discovered a deep garbage pit used by a tourist lodge. The neighboring troop took to forging daily in the dump, eating discarded salads, roast beef, and plum puddings. Somehow the baboons in his troop learned about the feasting at the garbage dump, and each morning, six of the most aggressive males headed for the dump, where they competed with fifty or sixty baboons from the other troop.

Not long after, tuberculosis broke out among the garbage-dump baboons. In humans, tuberculosis is a chronic disease the slowly wastes a person, but in nonhuman primates, TB is a wildfire, spreading rapidly, killing within a week. Sapolsky and Kenyan wildlife veterinarians found the source of the disease. The meat inspector at the lodge had been bribed to approve tubercular cows for slaughter, and their organs were discarded into the garbage pit and then consumed by the baboons.

All the males of Sapolsky's troop who had raided the dump died. Emotionally distraught by the devastation of his troop, Sapolsky switched to a study of a new troop miles away and did not return to his old troop for six years, and then only because he wanted to show his soon-to-be-wife the baboons of his youth. He was greeted by a pleasant surprise. He saw two adult males grooming each other, a rare event, except in this troop. The aggressive males had all died off, producing a ratio of 2:1 females to males, instead of the previous ratio of 1:1. The predominance of less combative females reduced substantially the aggression of the social life of the troop. Sapolsky also surmised that the adolescent males who "had grown up in typical baboon troops and then joined this one [had] adopted the style of low aggression and high affiliation. The troop's social culture" was transformed.[326]

When I read about the social plasticity of baboons, I was amazed. I, of course, knew that human life is extraordinarily varied and suspected that the radical changes in human culture are accidental, too, probably the unintended outcomes wrought by war, immigration, and technology. While the cultural resistance to change is always enormous, individual humans, unlike baboons, possess almost unlimited freedom to adopt a new way of life. Here, I am not speaking solely about saints and spiritual masters, but about ordinary people like you and me.

Animals Are Not Diminished Humans

Thus far, I have proceeded like a typical neuroscientist, primatologist, or biologist, ignoring the obvious — human beings possess language and no other animal does. Animals do have the shadow of language. Ants communicate through smell, bees through dance, and chimpanzees through sound and gesture. Animal communication is not a a diminished version of human language.

The prevailing view among scientists in the 1960s was that chimpanzees had to have the capacity for language because human beings are not unique, and evolution proceeds by small, incremental steps; hence no enormous gap can exist between chimpanzees and humans. The reason given that chimpanzees cannot speak a rudimentary version of English or German is that chimpanzees lack tongue flexibility and a resonant larynx; consequently, they cannot form the vowels and constants of human languages.

R. Allen Gardner and Beatrix Gardner, two University of Nevada psychologists, hit upon a brilliant idea — teach chimpanzees American Sign Language. The Gardners claimed their chimpanzee, Washoe, a female, used more than eighty-five signs after three years' training[327] and even coined new words; she signed "water bird," after seeing a swan.[328] Francine Patterson followed this sign language approach with Koko, a female gorilla.[329] Skeptics, however, were unconvinced.

To investigate if chimpanzees can truly learn sign language, psychologist Herbert Terrace organized an elaborate project.[330] At a cost of over $250,000, with four years' labor, Terrace and sixty other teachers attempted to teach American Sign Language to an infant male chimpanzee, nicknamed Nim Chimpsky, a playful taunt of Noam Chomsky, the founder of modern linguists, who insisted that language is innate and uniquely human. This project, the best documented of its kind to date, produced over forty video-tape hours of Nim signing and more than 2,000 teachers' reports.[331] Analyzing the

19,000 recorded signs produced by Nim, Terrace found "no evidence of lexical regularities," no sentences, no grammar.[332] Nim's numerous long strings of signs had no syntax, not even linking an adjective to a noun. Nim always combined signs in a series of repetitions with little new information and much redundancy. After his extensive analysis of Nim's signing, Terrace concluded, "Each instance of presumed grammatical competence could be explained adequately by simple nonlinguistic processes."[333]

In the 1960s, astrophysicist Carl Sagan and many other scientists hoped to read the diary of a chimpanzee to discover the interior life of one of our fellow creatures. What they would have read were pages and pages of Nim's longest combination of signs, for example, "Give orange me give eat orange me eat orange give me eat orange give me you," and such like "sentences," repeated again and again. Figure 15.2 shows a chimpanzee supposedly meditating upon such profound thoughts.

Figure 15.2. If we could listen to the meditative thoughts of a chimpanzee, we would hear, "Give orange me give eat orange me eat orange." C. H. Baum, *Sad Chimpanzee Thinking About His Life*, Shutterstock.

Terrace also analyzed the signing of Washoe and Koko and concluded that neither of them had learned sign language. The signing of both lacked syntax, and their lengthy strings of signs did not convey more meaning. Terrace pointed out that "Washoe may have simply been answering the question, *what that?*, by identifying correctly a body of water and a bird, in that order.

Before concluding that Washoe was relating the sign water to the sign bird, one must know whether she regularly placed an adjective (*water*) before, or after, a noun (*bird*). That cannot be decided on the basis of a single anecdote, no matter how compelling that anecdote may seem to an English-speaking observer."[334] One hundred percent of Koko's signs were prompted or asked for by her teachers, in marked contrast with young children's speech, which is more than eighty percent spontaneous and increases steadily in length, richness, and complexity.

The work of Terrace convinced linguists that animal communication lacks syntax and thus is not a truncated version of human language. After a thorough survey of the evidence that chimpanzees possess the ability to learn sign language, linguists Thomas Sebeok and Jean Umiker-Sebeok concluded, "Real breakthroughs in man-ape communication are still the stuff of fiction."[335] And Chomsky added, "It's about as likely that an ape will prove to have language ability as that there is an island somewhere with a species of flightless birds waiting for human beings to teach them to fly."[336] After Terrace's exhaustive, critical analysis, the entire field of teaching sign language to nonhuman primates collapsed.[337]

Sapolsky gives the reason why nonhuman primates cannot speak. Only the brain of *Homo sapiens* has Broca and Wernicke areas, the regions needed for language production and comprehension, respectively. The brains of the other primates, including chimpanzees, gorillas, bonobos, orangutans, and rhesus monkeys, have only the beginning of these structures, a mere cortical thickening.[338]

Because of the prevailing agenda of primatologists and evolutionary psychologists to establish that humans are "no better than brutes,"[339] the obvious must be stated. A comparison of Nim's signing — "give orange me give eat orange me eat orange" — with Terrace's writing — "each instance of [Nim's] presumed grammatical competence could be explained adequately by simple nonlinguistic processes" — is incontrovertible proof of the enormous gap that exists between us and nonhuman primates. Chimpanzees do not, and cannot, write peer-reviewed journal articles.

The life of nonhuman primates is locked into time, into the cycle of birth, growth, reproduction, senescence, and death. Their perceptual life, like ours, is of the here and now. Their memories, like ours, are of past particulars. We share with chimpanzees primate tool-making, nascent culture, and the desire to be consoled by touch, although we have perfected all these. The life we share with nonhuman primates and some other animals makes us a part of nature.

However, with language, we transcend space and time, even when we point and say "animal," for "animal" is a universal category defined by an organism that moves itself and has sense perception. We wonder about nature, about causes and effects, about how birds fly, why the trees turn color in the fall, and how bees find their way back to the hive. Through language, reason, and experiment, we attempt to discover the unchanging first principles of natural phenomena that apply universally, not merely to one or two particular instances in space and time. The principle of inertia states that matter resists a change in motion, not that one James Brown found it difficult to move his suitcase from his front porch to his car in Atlanta, Georgia on July 17, 1958. The principle of inertia applies to everyday objects moving at ordinary speeds, no matter where the place and what the time are. Our capacity to grasp universals and natural laws sets us apart from the other animals, and, in that sense, we are apart from nature.

How strange that *Homo sapiens*, a flash in the pan in cosmic history, can intellectually grasp the Big Bang and the Big Freeze, the beginning and the end of everything. Unlike zebras, kangaroos, and chimpanzees, human beings in some mysterious way transcend space and time; through science, philosophy, and art, we rise above nature. We live in time, yet touch the timeless. Compared with the other animals, we are special, wondrous creatures.

16 A Frog Tells Me Who I Am

I had the good fortune to be raised in the rural Midwest; otherwise, I think I would have been permanently lost. My Romanian heritage exposed the surface of American life with clarity, and my interest in other peoples of the world revealed the hidden structure of modern Western culture; yet, without my Huck Finn adventures into nature as a boy and without the long periods of solitude in the New Mexico high desert as a refugee from the Los Alamos Laboratory for the Destruction of Humankind, I would have remained bound by individualism, capitalism, and the nation-state. I know now there are other routes out of culture, but nature was my way.

I was ten years old when a frog gave me my ticket out of the prison of the modern Western way of thinking and feeling, and although at the time I did not know how to use the ticket, I held on to it. The O'Neill brothers and I had heard that you could eat frog legs. Our boyhood fantasy was to leave home for a summer and live entirely on our own in the woods. We had an unlimited supply of frogs in the surrounding ponds, and if frogs were edible, we could eat them and live in a lean-to made from pine-tree limbs.

Mrs. O'Neill told us that the French regard frog legs as a delicacy and that only in the finest restaurants in America could you order them. As a trial, we caught a half-dozen frogs, cut off their legs, skinned them, and asked Mrs. O'Neill to prepare them for us. She pan-fried the frog legs in butter, and much to our surprise, we found them delicious.

We set about catching frogs in a big-time way. The frogs rested on the banks of ponds or were half-hidden in ponds with only their eyes and nostrils above the water. Our hunting technique was crude. We used clubs made from fallen tree branches. We would sneak up on a frog and whop him one. Our hunting technique soon failed. The frogs leaped into the water and reappeared four or five feet from the bank, safe from our clubs. Patrick, the older O'Neill brother, hit upon a brilliant idea. He figured frogs ate flies, so we could catch frogs the way his father caught fish. So, we decided to go fly fishing for frogs. We "borrowed" three small artificial flies from Mr. O'Neill's fishing tackle box, and we fashioned frog poles out of long tree branches and out of the fishing line we had. Soon, we were hauling in frogs.

Bobby, the younger O'Neill brother, had a mishap. He hooked his fly on a tree limb that was half-submerged in the water. When he tugged to free the fly, the line snapped, and he lost one of his father's prize flies. Bobby was more concerned about being left out of the fun of catching frogs than he was about what punishment his father would mete out to him. He came up with a bizarre idea. He tore a fly-sized piece of fabric off the tail of the red shirt he was wearing and began to tie the cloth to the end of his fishing line. His older brother told him he was stupid if he thought any frog would mistake a small piece of cloth for a fly. Patrick and I could see that Bobby's "fly" bore no resemblance to a fly. Bobby ignored us and concentrated on tying the cloth to his line. When he began frogging again, he did not look confident. Patrick and I held our sides to keep them from splitting from laughter. Bobby dangled his "fly" in front of a frog. For a moment, nothing happened; then the frog unfurled its tongue and captured the "fly." The frog did not let go of its prize when Bobby swung his pole toward the bank. Patrick and I were amazed. We had read about monkeys refusing to let go of food, so the frog refusing to let go of its "fly" did not baffle us. What we could not understand, then or in our many subsequent discussions, was why the frog was fooled by such an obvious fake. Even the dumbest kid in the world could see that the piece of red shirt-tail did not look like a real fly.

It is strange how such an incident stuck in my mind. I must have sensed that nature had given me a profound lesson, although I was incapable of understanding it then. If only I had a mentor that I could have gone to and asked about the frog, I would have avoided spending years aimlessly searching for an answer to a question I could not even formulate.

Years later, in the physics library at the University of Michigan, I happened upon an issue of *Scientific American* that contained an article on frog vision. I was absolutely astounded by what I read, and I looked up the original journal article. "What the Frog's Eye Tells the Frog's Brain," was published in the most unlikely place, the *Proceedings of the Institute of Radio Engineers*, and was the collaborative effort of Lettvin, Maturana, McCulloch, and Pitts, a group of researchers from Massachusetts Institute of Technology.

The researchers inserted tiny electrodes into a living frog's optic nerve so that they could measure the electrical impulses traveling to the frog's brain. Using this technique, the researchers formed a good picture of what the frog sees and how. They found that when a small object is brought into the frog's field of vision and left immobile, the frog's eye sends electrical impulses to the brain for a few minutes but then ceases to do so. After a short time, the object is no longer there as far as the frog is concerned. The reason for this disappearance is that the frog's retina is designed to detect small moving objects. If a small object ceases to move in the frog's field of vision, the retina cancels it out of the frog's world. Furthermore, the retina's circuitry computes the velocity and trajectory of a small moving object so that the frog can aim its tongue ahead of where the object actually is.

The researchers reported that "the frog does not seem to see or, at any rate, is not concerned with the detail of stationary parts of the world around him. He will starve to death surrounded by food if it is not moving. His choice of food is determined only by size and movement. He will leap to capture any object the size of an insect or worm, providing it moves like one. He can be fooled easily not only by a bit of dangled meat but by any small object."[340] (See Figure 16.1.[341])

Figure 16.1. Non-moving insects and worms do not exist for a frog. It will starve to death, although surrounded by food.

A frog *cannot* see a fly as such; it sees small moving objects. The MIT researchers also discovered nerve fibers that respond to the net dimming of light. These specialized fibers alert the frog to the danger of a nearby large moving object. In effect, the frog's eye has only two categories: "my predator" and "my prey."

Similarly, ethologist Jacob von Uexküll, among the first to document the remarkable specificity of animal perception, discovered that a jackdaw is unable to see a grasshopper that is not moving: "A jackdaw simply does not know the shape of a motionless grasshopper and is so constituted that it can only apprehend the moving form. That would explain why so many insects feign death. If their motionless form simply does not exist in the field of vision of their enemies, then by shamming death they drop out of that world with absolute certainty and cannot be found even though searched for."[342]

The narrowness of animal perception can produce astounding results. Here are several of the hundreds of examples discovered by ethologists.

A deaf turkey hen will peck all her own chicks to death as soon as they are hatched. The distressed cheeping of the chicks is the only stimulus that inhibits the hen's natural aggression in defense of the nest. The cheeping alone evokes a maternal reaction in the hen. Without the cheeping, a chick is judged by instinct to be an enemy and is attacked. A hen with normal hearing will attack a realistic stuffed chick if it emits no sound and is pulled toward the nest by a string. Conversely, she will respond maternally to a stuffed weasel (the turkey's natural enemy) if it has a built-in speaker that produces the cheeping of a turkey chick.[343] Just as the frog cannot see a fly, the mother turkey cannot see its offspring!

Animal behavior is extraordinarily easy to misinterpret. For instance, a person observing a flock of jackdaws mobbing a cat that has captured one of their numbers might assume the birds had understood the situation and had chosen an appropriate course of action. Not so. Lorenz reports that on one occasion, when returning home from a swim, he was suddenly attacked by the normally friendly flock of jackdaws that nested on the roof of his home. When Lorenz withdrew from his pocket a black bathing suit, the birds screeched their sharp, metallic mobbing call and attacked Lorenz's hand to free the "captive bird." The flapping, black bathing suit triggered the jackdaw's mobbing instinct. Lorenz remarks, "Of all the reactions which, in the jackdaw, concern the recognition of an enemy, only one is innate: any living being that carries a black thing, dangling, or fluttering, becomes the object of a furious onslaught."[344] The jackdaws perceive black, they perceive flapping, but amazingly they cannot perceive bathing suit or even jackdaw. (See Figure 16.2.)

Primates, considered the most intelligent of animals, also do not perceive the *what* of things. Primatologist Wolfgang Kohler reports on the narrow perception of chimpanzees, "I tested them with some most primitive stuffed toys, on wooden frames, fastened on to a stand, and padded with straw sewn inside cloth covers, with black buttons for eyes. They were about forty centimeters in height and could perhaps be taken for caricatures of oxen and asses, though most drolly unnatural. It was totally impossible to get Sultan, who at that time could be led by the hand outside, near these small objects, which had so little real resemblance to any kind of animal. He went into paroxysms of terror, or threatened recklessly to bite my fingers, when I . . . tried to draw him towards the toy, as he struggled and strained backwards. One day I entered their room with one of these toys under my arm. Their reaction-times can be very short; in a moment a black cluster, consisting of the whole group

Figure 16.2. A jackdaw's mobbing instinct responds to "black, flapping." The bird will attack a cat carrying a blackbird or a hand holding bathing trunks. A jackdaw never attacks a cat when it is not carrying "black, flapping." (After Uexküll)

of chimpanzees, hung suspended from the farthest corner of the wire-roofing, each individual trying to thrust the others aside and bury his head deep in among them."[345]

How remarkable that the apes perceived the shape, size, color, and design of the stuffed toys but could not see what they *were* — harmless cloth and wood. Psychologist Patterson discovered the same thing while training her female gorilla: "Although Koko has never seen a real alligator, she is petrified of toothy stuffed or rubber facsimiles. . . . I have exploited Koko's irrational fear of this reptile by placing toy alligators in parts of the trailer I don't want her to touch."[346]

The great discovery of ethology is that animals do not perceive what things really are; an animal's perception is limited to a few key elements that will cause it to act. Uexküll summarizes the scientific study of animal perception with a powerful metaphor: An animal's world is not the world we see at all, but more closely resembles "a small, poorly furnished room."[347]

To picture the impoverished perceptual life of animals is extraordinarily difficult. We perceive the *what* and the *why* of things, substances and causes, not just black and flapping, but swimming suit and returning swimmer. Only extreme and rare pathology can cause a human being's perceptual life to approximate that of an animal. Oliver Sacks, a neurologist, reports that a neurological patient of his picked out key features of a scene, "a striking brightness, a color, a shape . . . but in no case did he get the scene-as-a-whole. . . . He had no sense of a landscape . . ."[348]

We know from experience that human sensory perception is not limited to rigid categories of utility. Entomologist Edward Wilson studies ants because he finds them fascinating, not because he wants to exterminate them. He says, "the primary aim [of naturalists] is to learn as much as possible about all aspects of the species that give them esthetic pleasure."[349] When she was seven, my daughter Brett Ann picked thistles with purple flowers and displayed them in a vase because they were beautiful. What is botany but the love of plants? We fall in love with whatever is beautiful and want to know more about it. Mathematician and theoretical physicist Henri Poincaré said, "The scientist does not study nature because it is useful to do so. He studies it because he takes pleasure in it; and he takes pleasure in it because it is beautiful. If nature were not beautiful, it would not be worth knowing and life would not be worth living."[350]

Animals have no relationship to things beyond utility. The bloodhound's sense of smell is acute enough to detect a person's unique scent from a five-week-old fingerprint, but the hound uses its expert nose to track animals, never to delight in the fragrance of a rose. A barn owl can distinguish objects in light one hundred times dimmer than the light human beings need to see anything, but the owl uses its keen night vision to sight rodents, never to study the yearly motion of Venus. An animal perceives things only under the aspect of what is useful or harmful to itself and hence ignores virtually all of nature.

Of all the natural creatures, only human beings can get beyond utilitarian needs. A frog cannot see the iridescent, filigreed wing of a fly, nor can the fly see the frog's glistening head and jet-black eyes. A ten-year-old boy seated on the bank of the pond can take in the frog and the fly, can see the puffy white clouds racing across the blue sky, and can feel the warm spring breeze. Without the presence of a human being, the scene does not exist.

Every animal is trapped within the narrow world of utilitarian desire, unaware of even its own beauty. The tiger does not know what a wonderful thing a tiger is; only a human person knows that. What characterizes human life in contrast to animal life is that a human person can get outside himself or herself through love.

If you want to experience the human way of life, go outside at night and look at the stars, or in summer, pick up a dandelion and look at it, or gaze into the eyes of the next person you see. Only a human person can fall in love with the other; only a human person is open to all existence. Who are we? We are lovers. Each person is, as it were, the eyes and ears of nature. Instead of inhabiting "a small, poorly furnished room," each human person through love can be connected to all that is.

We Are Spiritual Beings

Of all the natural creatures, only human beings can grasp a whole. The study of animal perception re-discovered the spiritual nature of *Homo sapiens — the capacity to be connected to all that is,* a fundamental principle of every wisdom tradition.

Ancient Greek: "The human soul is, fundamentally, everything that is."[351]

Hindu: "Thou are that."[352]

Christian: "Every other being takes only a limited part of being whereas the spiritual soul is capable of grasping the whole of being."[353]

Jewish: "At opposite poles, both man and God encompass within their being the entire cosmos. What exists seminally in God unfolds and develops in man."[354]

Islamic: "Who knows his soul knows his Lord."[355]

Chinese: "He who cultivates the Tao is one with the Tao."[356]

Native American: "To walk the path of beauty, you must connect to all things, take them seriously, with reverence."[357]

17 Emotional Habits That Accord with Our Spiritual Nature

If I was initially put off by Tocqueville pointing out that democratic culture programmed my thinking, then the corollary I drew that the same culture instilled emotional habits seemed even more outrageous. And, for a good reason. The emotions appear to be the core of our being. Our perceptions are of a common world available to all. Our thoughts are easily made public and can be shared by others. But our emotions seem radically private because no one can "feel what I feel." Most of us consider the most private aspect of ourselves to be who we are. Like nearly all Americans, I believed that a unique emotional profile defined me as an individual and provided the answer to the question "Who am I?" If I am occasionally moody or often lonely, then that is just the way I am. If I am quick to anger or slow to respond, that is the kind of person I am. However, the belief that our emotions are unique is false because isolated individuals share the same emotional profile. If we are isolated from others, we see only our own needs and desires. Nothing else is apparent to us. Our isolation leads us to see the world and other persons only in terms of our wants. When others frustrate the fulfillment of our desires, we become angry; when others have what we lack, we become envious. When we cannot get what we desire, we feel sorry for ourselves. Our isolation from others, then, predisposes us to such negative emotions as anger, envy, and self-pity.

Circumstances modify this emotional profile; nevertheless, all isolated individuals have a common profile. Consequently, the emotional responses of Americans are predictable. Drive any freeway in Massachusetts or California, and you see raw anger, the honking of horns, and the shaking of fists. If you want to see how capitalism fosters envy, watch "Billionaire Rich Lifestyle" on YouTube. If you want to hear an expression of self-pity, turn on any country-and-western radio station in any city in the United States.

But equality tempers the negative emotions of the isolated individual. When we Americans see hardships and suffering, we easily imagine ourselves in those of need. We generously support philanthropic enterprises; in times of disaster, such as hurricane Katrina or the attacks of 9/11, the deeper the misery of others, the more our hearts pour out words of sympathy, and our hands write larger checks. Many of us travel great distances and endure hardships ourselves to aid disaster victims. The source of such noble action is democratic equality that engenders generosity in the American character.

Anger, envy, and self-pity are such a part of everyday American life that we take the intensity with which we feel these emotions as natural. But the Lakota or the Eskimo rarely experience these emotions. Eskimos, for example, speak of the white man's "world where people are always loud and angry."[358] In Eskimo culture showing anger to someone's face is not acceptable. Anthropologist Jean Biggs was ostracized from her adoptive Utku Eskimo community for several weeks because of her single outburst of anger against white fishermen who she thought were taking advantage of the Eskimos. Anger frightens and sickens Eskimos, for it destroys their web of human relationships. The rarity of anger in Eskimo culture harmonizes with the need for group solidarity. The prevalence of anger in American life suits a social and economic system where isolated, autonomous individuals compete for prizes, prestige, and material goods.

How Emotional Habits Can Be Changed

Neither intellectual insight nor verbal command alone will change an emotional habit. When sad or lonely, we may tell ourselves to cheer up; yet, the gloom remains, or if it does depart, it soon returns. We may realize that we grew up isolated from other persons and are thus distrustful of human relations; yet, we seem unable to break through the barrier that separates us from others. Most of us know that we have the power to choose what we think and how we act but mistakenly believe we cannot change how we feel. What we fail to grasp is that habits are formed through repeated action and that a new habit can *only* be acquired through the performance of a different action.

Consider a piano virtuoso, such as Alicia de Larrocha. In playing a passage from a Mozart cadenza, she does not stop to think which finger will play middle C. Her fingers move out of habit, in the same way, that a keyboard operator types the word "the" without thinking. But suppose when Miss Larrocha was a beginning student, she learned to misplay the passage. Then, to replace a bad habit of playing with a good one, she had to struggle to play the passage correctly over and over, first slowly, then up to tempo, until a new habit was acquired.

An undesired emotional habit can be eliminated in two ways, by direct or indirect action. Since an emotion moves a person toward or away from an object, one way to eliminate an emotional habit is for a person to use willpower to perform an action that opposes the tendency of the emotional habit.

For example, when I was ten years old, Ed and Frank Fleck, both in their twenties, lived directly across the street. To my ten-year-old mind, the Fleck brothers knew everything there was to know. They built boats, repaired cars, knew how television sets worked, and even had their own amateur radio station.

The Fleck brothers had a mammoth workshop. I often went down into their damp basement to borrow tools. One day when I flipped the toggle switch to turn off the overhead fluorescent lights, my right hand was suddenly paralyzed, and my knees buckled. I fell to the damp floor, and my heart did weird things. I managed to stand up, but my knees shook. I knew I had narrowly escaped death by electrocution.

Later, Ed, the older brother, took the switch apart and explained to me how it had failed. He repaired the faulty switch, but I refused to touch it with my bare hand. Several days later, when I went to borrow a vacuum tube voltmeter, I used a piece of dry wood to turn the lights on and off, even though I knew I would not receive an electric shock if I touched the switch. My reason told me, "The switch has been repaired and is harmless, now," and my emotions told me, "If you touch that switch, you'll hit the floor again."

My emotions always prevailed until several months later, Ed asked me to turn off the fluorescent lights in the basement. I was momentarily paralyzed, for I was embarrassed to have him see me use a dry stick to throw the switch and afraid to touch the switch with my hand. Only by a supreme act of the will did I switch off the lights with my hand. After that, I never feared touching the switch.

To use sheer willpower to eliminate a bad emotional habit is generally difficult and unpleasant. A better strategy is to use indirect action. Surprisingly, if a person focuses his or her awareness on the *physical* component of an emotion, the emotion spontaneously weakens or disappears altogether. Suppose a man has formed the bad habit of becoming angered by the slightest inconveniences, or worse yet, just waits for something to happen so he can become angry. He can change this habit by focusing his awareness on the tightness in his chest every time he gets angry. Over time, his anger will become less intense and less frequent, and eventually his bad emotional habit will be gone. But if the man focuses his awareness on the emotion itself, his anger will intensify and become more deeply established.

Another indirect way to eliminate an undesired emotional habit is through gradual desensitization. Many people, for instance, have a fear that is irrational even to them, such as the fear of being away from home, flying, or public speaking. What keeps a fear alive is avoidance of what is feared. Avoidance produces an immediate sense of relief but entrenches the fear. To overcome a fear, a person must face what he or she fears, which takes real courage, especially if the fear is a long-standing one. Most people find it easier to gradually face their fear, often with the help of a friend.

Over thirty percent of Americans report that public speaking is their number one fear, ahead of the dark, heights, loneliness, sickness, and even death. For a telephone linesman, an assembly-line worker, or a lumberjack, the fear of public speaking is no more than an occasional nuisance, but for a freshman college student enrolled in a seminar in English literature such a fear can be disastrous. Consider a young woman who fears being called upon by the professor and tries to hide by sitting at the corner of the seminar table. When she sees connections in a novel that the other members of the seminar miss, she ardently desires to contribute her insights to the discussion but cannot. Skipped heartbeats, a blushing face, and light-headedness prevent her from speaking. Her inability to speak in the seminar leads to embarrassment, lack of confidence, anger at herself, and increased fear.

To overcome her fear, she could start with an easy task, say, forcing herself in the seminar to ask one simple question, such as "I'm sorry. What did you say?" This first step probably would make her slightly anxious but not frighten her so much that she could not do it. Once comfortable with asking simple questions, she could offer her opinion or insight in terms of a question: "Is Aaron's relationship with his mother important here?" Later, she could have a friend in the seminar ask her to repeat her question, and, in this way, she would begin to engage in public discourse. Through such gradual desensitization, her fear of public speaking will weaken. With courage, she will overcome the inevitable setbacks, and with persistence, she will continue, for years if necessary, to break down a bad emotional habit.

A desired emotional habit can be acquired by using the principle that a person becomes what he or she does. Alicia de Larrocha became an excellent pianist by playing the piano excellently. This may seem circular, but it is not. A person becomes courageous by performing courageous acts or generous by performing generous acts. Since we become the way we act, a person can become a different kind of person by acting *as if* he or she were already that kind of person.

Anthropologist Ashley Montagu gives his own life as an example. He confesses that he was "brought up as a stiff, stuffed-shirt Englishman who considered that any exhibition of emotion was low class. To be very cutting in one's wit no matter how unpleasant it was, how denigrating it was to another person, was correct behavior."[359] Montagu says he changed from a nasty, hostile, aggressive creature simply by acting as if he were a loving human being. He recommends, "If you're not yet a loving human being, what you have to do in order to change is begin to act 'as if' you were by demonstrative acts, by communicating to others, by throwing your arms around them, by taking them by the hand, by putting an arm around their shoulders. It's enormously important to remember that 'as if.' You behave 'as if' you were a loving human being. If you go on behaving 'as if' you were a loving human being, one day you'll wake up and find you've become what you've been doing."

I learned how to acquire new emotional habits from Aristotle, not Montagu. I first read Aristotle's *Nicomachean Ethics* when I was twenty-six and was skeptical of the philosopher's contention that a person becomes generous or courageous by repeatedly performing generous or courageous acts until a habit is established. I decided to test what I had read. At the time, I was geeky, socially awkward, and withdrawn, not unlike many of my physicist colleagues, all seemingly afflicted with Asperger's disorder. If Aristotle were correct, then I would not necessarily be condemned to an emotional life determined by the accidents of childhood, by my neglectful upbringing that made me shun human contact and made me more at home with machines, electronics, and physics apparatuses. In the hope of becoming socially adept, I forced myself to go up to anyone I felt uncomfortable around and initiate a conversation. While standing in a grocery-store line, I talked to strangers, at times asking about their lives. Within six months, I enjoyed the company of other humans, took a genuine interest in them, and delighted in the diverse stories I heard. Nowadays, my friends cannot believe that I was once a mousy introvert, seldom seen and never heard.

Choosing Emotional Habits

Human life is extraordinarily difficult, for what is good for us is so deeply hidden. Except for *Homo sapiens*, instinct determines what is good for an animal and how to achieve it. A chimpanzee, for example, "knows" by nature what to eat, what to flee from, how to live in a group, and when to breed. Instinct directs each higher animal to have the right emotion at the right time, to the right degree, and for the right purpose.

If only human life were so simple. We must discover our deepest nature — *the capacity to be connected to all that is.* To further complicate life for us, unlike chimpanzees and all the other animals, we humans have two ways of appraising what is good for us — the mind and the emotions — and neither appraisal is determined by nature; even worse, the mind and the emotions often give opposite appraisals of what is good for us. A person may know that he should lose weight, but he likes crusty French bread, camembert cheese, and red wine; to further his career, he should volunteer to give the next public presentation at work, but the thought of speaking in front of a group terrifies him; he should join a gym or a reading group to make new friends, but lacks hope the effort will pay off.

Since nature does not prescribe a fixed way of life for us, we have extraordinary freedom. We can become hunters like the lions, carpenters like the beavers, or musicians like the birds. Every animal activity we raise to a new level. Yet, because of our freedom and undetermined nature, we become confused and acquire faulty intellectual and emotional habits frequently at cross-purposes.

Are there any emotional habits that everyone should strive to acquire or attempt to avoid? At first, the answer seems "no," since the human person is so unfinished by nature, and every culture constructs a different "I." But one universal, ordering principle for human life does exist: Every person has the capacity to be connected to all that is. Thus, any habit that disconnects a person from other people should be avoided. For instance, a person quick to anger will not only anger those around him or her but will be unable to judge irksome situations accurately. A short-tempered person often gets angry with the wrong people under the wrong circumstances and afterward may feel regret.

Just as the short-tempered person is avoided by others, so too is the grouch. No one desires to be around a person enveloped in gloom and doom or to associate with a man or woman who sees only shadows and what is wrong with others. The constant complainer refuses to put up with anything, no matter how trivial, and as a result is quarrelsome.

If a person cannot share, he or she is cut off from others; so, clearly, stinginess and greediness are to be avoided. Stealing and cheating are worse, for the thief and the cheat cannot disclose their activities for all to see and consequently have cut themselves off from humankind. Similarly, the known liar will not be listened to by others. The liar deprives himself or herself of the full use of language and reason, the two faculties that differentiate human beings from the animals.

Thus, short-temperedness, quarrelsomeness, stinginess, greediness, and deceitfulness cut us off from others; but friendliness, a cheerful disposition, generosity, and truthfulness connect us to others.

If truthfulness is extended beyond truth-telling to include the capacity to see things exactly as they are, free from subjective distortions, then truthfulness also means to see oneself exactly for what one is, neither more nor less. Such self-awareness is humility. Objective sight reveals that underneath the faults and weaknesses of one's neighbor lies suffering and a profound unknown. Compassion flows from seeing that one's neighbor is essentially no different from oneself.

Ignorance, greed, and anger, what Buddha called the three poisons, are the obstacles that stand in the way of a person becoming connected to all that is. If we wish to become who we truly are by nature, we must strive to be truthful, selfless, and compassionate.

18 We Are Meant to Love and to Be Loved

At the core of human life is love, which is obscure to us because in English love is an ambiguous word that can mean sexual attraction, affection for another, or even a strong like, say for a particular sport or food. In Koine Greek, on the other hand, love is divided into four kinds, *storgē*, *erōs*, *philía*, and *agápē*.[360] *Storgē*, often called familial love, is a natural affection that arises from the familiarity of persons, such as two women who daily sit next to each other on a commuter bus. The most intense form of *storgē* is that of a parent for an offspring. However, *storgē* is so broad that it even refers to the relationship between pets and their owners. In American English, the word that corresponds most closely to *storgē* is "affection."

Erōs

Erōs is an intense, passionate desire to be joined to another person, to beauty, to truth, to God, or to any good outside of oneself. We were born radically incomplete. When we emerged from the womb, we sought a human face and listened for a soprano voice. Our very first experience in life was connecting ourselves to another person. The first word spoken about us was either "boy" or "girl," a word that drew attention to our anatomy that exhibited that we were physically incomplete but had the latent desire for a profound union with another person. As a baby, we were a lovable bundle of *erōs*, the natural desire for full existence. This self-love is not a selfish love, although like any love, it can become selfish.

Say as an adolescent, we fell in love with Mozart's music. Such *erōs* developed our hearing, attuned our soul to beauty, and conformed our emotions to a rational structure. When we heard Mitsuko Uchida play K. 545, the *Sonata Facile*[361], we also fell in love with Ms. Uchida, the embodiment of musical perfection; such love is often called earned or deserved love; we hoped in some small way to return the great good she had given us.

In the ancient world, two lovers under the sway of *erōs* strive to gain an intimate knowledge of everything that pertains to the beloved, to penetrate the other's soul. Such lovers seek to possess each other perfectly by entering the heart of the other. At times, two lovers wish to be united into one, a union that would destroy one or both. Aristophanes claimed that if the god Hephaestus came to a pair of lovers and offered to weld them together, no lovers would refuse to be merged into an utter oneness.[362]

In Modernity, *erōs* is usually taken to be romantic love as depicted by movies, magazine ads, popular songs, and romance novels. Psychoanalyst Erich Fromm describes how this intense, all-consuming love happens: "If two people who have been strangers, as all of us are, suddenly let the wall between them break down, and feel close, feel one, this moment of oneness is one of the most exhilarating, most exciting experiences of life. It is all the more wonderful and miraculous for persons who have been shut off, isolated, without love."[363]

Fromm points out that in Modernity because of individualism, "One can often find two people 'in love' with each other who feel no love for anybody else. Their love is, in fact, an egotism à deux; they are two people who identify themselves with each other, and who solve the problem of separateness by enlarging the single individual into two."[364] The love of such lovers is exclusive; friends, family, and previous interests must be brought within the sphere of their love or abandoned outside. In practice, little can be incorporated into their love; so, in effect, they are separated from the rest of humankind; in their view, this makes their love more magical.

As I or any older adult can testify to personally, almost no one escapes the madness of being crazy in love. When a young couple falls in love, they are primarily in love with the pleasurable emotion produced by the other person. Such lovers wish to be together all day and to live together forever. When apart, life is flat, and each lover dreams of the other. Such a love frequently dissolves, often painfully, because ultimately, when push comes to shove, it becomes clear that each lover is pursuing his or her own ends and wishes to receive more than to give.

Counterfeit Love

In high school, I must have appeared like an exchange student from another planet. My hair was never combed; I sported an Afro years before Afro-Americans introduced the hairdo into mainstream America. My unmatched clothes resulted from gifts from various relatives. In the classroom, my demeanor alternated between brooding silence and impassioned outbursts about tangential topics. Given such weirdness, you would think not one girl would go near me. On the contrary, many girls found me a mystery, the only alternative in our high school to the fair-haired, all-American jock. Girls pursued me, and always in the same way. A girl would let me know through the grapevine that she wanted me to ask her for a date, and I willingly obliged. The only girl I knew in high school who did not work through the grapevine was Rita Willihnganz. She was a year behind me, and I only knew her by name. One day she walked up to me and bluntly asked, "Why don't you ask me out?" I did, and for years afterward my life was never the same.

I fell head over heels in love with Rita. I found her turned-up nose, her blond, bobbed hair, and her blue eyes divine. She was like no other girl I had met. When Rita looked at me, her innocent face told me that at that moment, I was the only person in existence. Her openness said that I had nothing to fear; I could be myself, and no matter what that self was, she would find it marvelous. Her inquiring eyes told me, "You are a mystery that I want to know."

When the two of us were together, the world orbited around us. Our magical love gave significance to foreign movies, summer stock theater, and restaurants. Without our love, the world would be nothing.

Everything in my life had become perfect. The world was enchanted. At the beach, beads of water hugged the delicate curves of Rita's body. The bright afternoon sun loved her golden hair. At night, we stood in the moonlight, in front of her house, and kissed goodnight. In the background were silhouettes of trees, and if music had begun to play, I would not have been surprised.

I became one of those mad lovers who are subjects for comedies. I took up the ukulele, and at night as our canoe glided across the still waters of Green Lake, I serenaded my love. I used the boxing lessons that I received as a boy from Mr. O'Neill to defend my love's honor. The high school boys to show their manliness often made disparaging remarks about all the girls, except Rita. I would go insane if any boy even suggested that Rita might have a blemish. If Rita's father had forbidden me to see his daughter, I swear I would have kidnapped her, for I could not live without her.

My falling in love was completely predictable, although at the time, I had not the slightest idea what was happening to me. In my ignorance, I thought I finally grasped human happiness, for I knew I was in love. The feelings I experienced were overwhelming. Every time I was with Rita or even thought of her, I was swept away on a sea of joyous emotion.

Our love was exclusive; friends, family, and previous interests had to be brought within the sphere of our love or abandoned outside. In practice, little could be incorporated into our love; so, in effect, Rita and I were separated from the rest of humankind, and this made our love all the more wonderful. Since no other person had ever made me feel like this, I knew that Rita was the love of my life, the only person in all human history that I could love, and I was the only person she could love. This was true love; we were fated for each other. On our very first date, I immediately recognized that I had met the person intended for me.

Rita gazed into my eyes, her fingers gently touched the back of my hand, and suddenly the wall between me and another person was broken down. Never before had I felt this oneness with another, and it was the most exhilarating experience of my life. Miraculously, love had healed me; I was no longer isolated, lonely, or empty inside.

I not only believed the myth of romantic love, I *lived* it; and I lived it more intensely than any fool I've known. But how could I know that the love that I experienced was a counterfeit love? Isolated from others, I was profoundly lonely and craved human companionship. Before I met Rita, I felt desolate and at times empty inside and haunted by the fear that I was not lovable. To numb the pain of my sixteen-year-old existence, I started to drink my father's Seagram's Seven and listen to Billie Holiday records. I didn't think anybody in the world really cared about me or what happened to me. I knew that I could not force any person to love me, so I passively waited and hoped for some young woman to come along and love me.

I now know that loneliness is the constant companion of every American. The more the self is emphasized, the more a person becomes isolated. The more I was cut off from others, the more desperately I sought love. As it had for countless others before me, love stepped in as a remedy for my loneliness.

Looking back, I now see that I did not love Rita, for I hadn't a clue about what it meant to love another person. I loved the emotions that Rita produced within me. I loved how she made me feel good about myself. While I thought I loved Rita, in fact, I fulfilled my own emotional needs. Just like an infant, I could not love; I could only be loved. In truth, through Rita, I loved myself.

Despite my illusions of oneness with another, I was still isolated. I suspect that Rita was also isolated and not in love with me. She was in love with romantic love and assigned me to play the romantic lead in the script she imagined. Rita fashioned her script from bits and pieces of mass culture. Unbeknownst to me, she lifted words from Broadway musicals and stole gestures from the movies. Only recently did I discover that the special terms of endearment that she taught me and that we whispered to each other were taken from the musical *Fantasticks*. Our love was a dual egoism, held together by the exchange of needs: She needed a romantic lead, and I needed emotional comfort.

Philía

Philía is usually translated as friendship, although the ancient understanding of *philia* is much wider, for *philia* holds the members of any association together, whether it be a city-state, a business partnership, or even a buyer and a seller. *Philia* between buyer and seller should exist before, during, and after an exchange of material goods. In capitalism, the relationship between buyer and seller is contractual.

My parents ran their grocery store in rural Michigan according to the Romanian peasant understanding of social life. Our store served the rural community of Union Lake, forty miles west of Detroit. Most businessmen nowadays believe the purpose of a business is to make money, and consequently they readily destroy community.

Every Christmas Eve, my father loaded the store's panel truck with bags of groceries which he delivered to poor people in our neighborhood. During strikes at General Motors or Ford Motor, my father carried strikers on the book, often for five or six weeks. My parents told me repeatedly that our store was there to serve others and that we would do well by doing good; my father added that the other guy often needs help, even if only in small ways.

In his *Nichomachean Ethics*, Aristotle argues that we love what is good, pleasurable, or useful about another person, and thus there are three kinds of friendship.[365] When the motive of friendship is usefulness, we do not feel affection for another as such but seek to fulfill the desire for some material good through them. These friendships are not always morally reprehensible. Many friendships of utility, for instance, are founded on the exchange of favors. These relations are common among neighbors and acquaintances and include such everyday associations as carpools and food co-ops. In the work-place, where utility generally reigns, many people, because of ambition, wish to receive affection from superiors and use flattery, pretending to be a friend in an inferior position. In the vernacular, ass-kissing is a common way to advance a career since most men and women love flattery, for they wish to be loved rather than to love.

Friendship based on usefulness is subject to complaints and rapid dis-solution. The sharing of money, power, and honor invariably leads to disputes. The desire for more is insatiable, and the feeling of being treated unfairly is seldom absent. Aristotle, in his realistic manner, observed over 2,300 years ago, "Most people wish to be recipients of good deeds, but avoid perform-ing them, because they are unprofitable."[366] Some things in human life never change.

Friendships based on pleasure are not that different from friendships of utility. We love witty people not for what they are but for the pleasure they give us. Children call one another friends because of the pleasure they have when playing together; yet, such childhood friendships advance human growth and development.

When he was four years old, my grandson Yasu told me he had "lots of friends." He played more intensely and more imaginatively when one or two of his friends were present. I watched one of his friends imitate Yasu jumping from a stool to the floor. The two friends soon developed a game in which the two boys took turns standing on one leg on the stool, shouting something about pirates, and then jumping to the floor. In their play, Yasu and his friend developed social skills, physical dexterity, and imagination.

The friendships of adolescence are based on pleasure, too. The lives of young adults are guided by emotion, and thus, for the most part, their friend-ships are short-lived.

In the third kind of friendship, based on the good, a common life is shared. Whatever makes life desirable is pursued together. Some friends en-gage in sports together, others perform music, yet others pursue social justice. Unlike *erōs* where lovers gaze at each other, friends under the power of *philía* gaze at a third thing outside of themselves. What friends love most in life is what joins them together; their mutual love enhances their love of the third thing that binds them together. Such friends feel pleasure and pain from the same things, judge the same way, and understand the same things — in a sense, they are one soul, and such unity of souls, in itself, is pleasurable.

In the truest, most long-lasting friendship, a friend acts for his friend's good for the friend's sake, as if his friend were another self. In this highest form of friendship, we love our friends in the same way we love ourselves. Aristotle defines the highest form of friendship as the love that wishes another everything we think good, and moreover for the other's sake, not for our own, and to bring these goods things about for the other, as far as we can.[367]

Human life, of course, is messy and never fits the neat compartments of philosophical analysis. Consider a teacher, say of philosophy, physics, or poetry. The teacher needs the student's tuition money, and the student needs a good grade from the teacher for a diploma and a good job. Being human, especially when young, a teacher desires adulation from his students and sees teaching as performance, after which students praise how smart and talented he is. A teacher and his students find a public display of witty erudition mutually pleasurable. Some teachers pursue the truth with one or two of their students and, in this way, develop friendships based on the good. Teachers often have friendships with their students based simultaneously on the useful, the pleasurable, and the good.

Agápē

In the New Testament, the Gospel of Love, the Greek word *erōs* does not occur once, while *agápē*, infrequently used in ancient Greek, occurs 116 times and stands for a new understanding of love. In the King James Version of the Bible, *agápē* is translated as charity, a word that now means to most people giving handouts to the homeless or contributing to United Way. In the Revised Standard Version of the Bible, *agápē* is translated as love, an ambiguous word in English that can mean romantic love or even a strong like, say for major league baseball or Italian food.

In the Christian tradition, *agápē* means God's selfless love for human persons, a love that cannot be earned and excludes no one.[368] Such love gives and expects nothing in return. Since no English translation accurately renders how *agápē* is used in the New Testament, probably the best recourse is for the reader to stop, ignore the proper Greek, which only a minority of us know, and try substituting *agápē* and *agápēd* for "love" and "loved." For example,

> John 13:34: A new commandment I give to you, that you
> love [*agápē*] one another; even as I have loved [*agápēd*] you,
> that you also love [*agápē*] one another.

Or,

Matthew 5:43-46: You have heard that it was said, 'You shall love [*agápē*] your neighbor and hate your enemy.' But I say to you, [*agápē*] your enemies and pray for those who persecute you, so that you may be sons of your Father who is in heaven; for he makes his sun rise on the evil and on the good, and sends rain on the just and on the unjust.

Despite the butchering of the Greek, this substitution shows that Jesus is calling upon us to practice a new love, to love the way God does, to love our neighbors and even ourselves selflessly and unconditionally, without desiring a reward, without wanting something in return. For we moderns, *agápē* calls us to abandon the narrow self-love of individualism and to embrace the expansive love rooted in our spiritual nature — *the capacity to be connected to all that is* — to love the world and its inhabitants the way God does.

But as we have seen in our discussion of *erōs*, all human love begins with self-love, and thus many see Jesus' call for a new love as humanly impossible. Thomas Aquinas argues that "*agápē* itself surpasses our natural facilities" and that we can only exceed human nature by the Holy Spirit infusing *agápē* into our souls.[369]

Hence, for Aquinas, *agápē* is beyond ordinary mortals like you and me; not one of us will ever be a Mother Theresa ministering to the dying homeless in Calcutta. However, if we relax the selfless characteristic of *agápē* and focus on the unconditional aspect, we will see that *agápē* is not beyond us.

Unconditional Love

As we have seen, the very first experience in any baby's life is connecting himself to another person. Within days, he can distinguish between his mother and others by her looks, voice, and smell. The mother, on her part, desires to cradle her infant, to soothe him when he cries, to keep him warm and protected. The infant shares an interior life with the mother. If she becomes startled or anxious, the baby becomes frightened and cries. If she coos, the infant coos back. A mother and her infant often play the cooing game, each taking pleasure in sharing emotion.

Indian psychoanalyst Sudhir Kakar reports that "well up to the fifth year, if not longer, it is customary for Indian children to sleep by their mother's side at night. . . . Constantly held, cuddled, crooned, and talked to . . . the young child has come to experience his core self as lovable: 'I am lovable, for I am loved.' Infancy has provided him with a secure base from which to explore his environment with confidence."[370]

As Kakar points out, unconditional love is the foundation that supports the further growth and development of the child. For the young child, everything pivots around the mother or the continuous caregiver; she is the entire world. John Bowlby, a psychiatrist and the recognized authority on the emotional attachment of children to their caregivers, gives the example of "a healthy child whose mother is resting on a garden seat will make a series of excursions away from her, each time returning to her before making the next excursion."[371] The little excursions the child makes beyond her are rooted in the confidence that she will always be there for protection and comfort. The mother's love provides the child with a secure base to explore the world.

Fromm observes that "unconditional love corresponds to one of the deepest longings, not only of the child, but of every human being."[372] No one wants to be loved only for economic success or artistic achievement. Earned love always leaves the lingering doubt that one is not loved for oneself but for the status and pleasure that accrues to the "lover." A doubt about earned love often becomes the nagging fear that one is not loved at all but used.[373]

On rare occasions, we witness that leap from *erōs* to *agápē* in a person's love of another. Consider Lasandra Carter, a Black woman in Detroit, probably in her late thirties, poorly educated, and pregnant with Giovanni. In utero, he was found to have major genetic abnormalities guaranteed to drastically shorten his life after birth.

Before his delivery, Lasandra appeared depressed and lifeless. After the birth of Giovanni, Lasandra seemed happy and in love with tiny Giovanni. "Everything is so big on him; he's such an itty-bitty guy!" she exclaimed. She took off one of his tiny socks and said, "He has a little clubfoot, such a cutie," and then caressed his clubfoot. Lasandra showed unconditional love for Giovanni: "We have some time with him, but not a lot. The only thing I can do is love him until the time comes."

Undoubtedly, Lasandra never heard of *erōs* and *agápē*; yet, her heart had, for the video of her loving Giovanni shows the miracle of an ordinary person leaping from *erōs* to *agápē*, from being concerned with only herself to her selfless love for another.[374]

If we carefully observe the social life around us, we see the meanspirited, the ambitious, and the selfish, but we also see numerous physicians, nurses, teachers, and caregivers making the miraculous leap from *erōs* to *agápē*.

To love well, we must be loved unconditionally first. Psychologist René Spitz discovered through the study of hospitalized children that a child's very first bond with another person is the basis for the later development of human love and friendship.[375] A child under two years of age, if deprived of a single person's continuous care for three months or more, develops emotional trauma that may result in death, even though the child is provided with perfectly adequate food, shelter, hygiene, and medication by a succession of compassionate nurses. In such circumstances, no one is exclusively responsible for loving the child, so she cannot form an attachment to another person.

Spitz recounts the suffering of one baby girl deprived of her mother: "She lay immobile in her crib; when approached she did not lift her shoulders, barely her head, to look at the observer with an expression of profound suffering sometimes seen in sick animals. As soon as the observer started to speak to her or to touch her, she began to weep. This was unlike the usual crying of babies, which is accompanied by a certain amount of unpleasure vocalization, and sometimes screaming. Instead she wept soundlessly, tears running down her face. Speaking to her in soft comforting tones only resulted in more intense weeping, intermingled with moans and sobs, shaking her whole body. This reaction deepened in the ensuing two months. It was more and more difficult to make contact with the child. Seven weeks later, it took us almost an hour to establish contact with her. During this period, she lost weight and developed a serious eating disturbance; she had difficulty in taking food and in keeping it down."[376]

If the separation continues, the child shows a rapid decline in mental and motor development, eventually being unable to sit, stand, walk, or talk despite the best of institutional care. In the extreme case, when love is totally absent, or nearly so, the child simply dies, or if it survives, its emotional life is permanently damaged. What a child, and even an adult, needs over and beyond food, shelter, and other physical necessities is the love of another person.

Spitz also discovered that when a child experiences a mother or a primary caregiver as a source of both intense pain and comfort, all the child's emotions are blurred, and its capacity for friendship is severely diminished. A child severely deficient in love is not interested in his or her toys and is prone to violence in later life. An empty, uninterested facial expression is a characteristic of a child lacking love. Many a child's life has been saved from ruin by the sustained, unconditional love of a grandmother, an aunt, or a nanny. If a mother or continuous caregiver showers the baby with gratuitous love, the infant feels, "I am wonderful, just because I am." The child learns to love itself the way the mother or caregiver loves him or her. The young child then extends this self-love to a love of the world. The child feels, "It's good to be alive; it's good to be surrounded by such good things." With unconditional love, a child learns to trust life.

A child nurtured and protected by love can as an adult suffer the most outrageous misfortunes and still believe she and the world are fundamentally good. If success is measured by human relations and friendship, not wealth and career achievement, then the kind of love a child receives is a better predictor of her course in life than environment, IQ tests, or genes.

No One Is Condemned to a Loveless Life

If a person never received unconditional love as a child, he or she is not condemned to a loveless life. Consider Godfrey Camille, a participant in the

Harvard Study of Adult Development.[377] He had the bleakest childhood in the Study; his parents were upper-middle class, socially isolated, incompetent, and pathologically suspicious. Later in life, Camille confessed that he neither liked nor respected his parents.

As a college student, Camille sought love and care by frequent visits to the infirmary. In his junior year, the usually sympathetic physician contemptuously dismissed him with the evaluation, "This boy is turning into a regular psychoneurotic."

At his ten-year personality assessment, the Study consensus was that Camille was "not fitted for the practice of medicine." Nevertheless, he went to medical school, and shortly after graduation, he attempted suicide. He was overwhelmed by the thought that he would have to take care of other people when he could not take of his own self.

At the age of thirty-five, Camille had a life-changing experience. He was hospitalized for fourteen months with pulmonary tuberculosis. His time in the hospital was like a religious rebirth: "Someone with a capital 'S' cared for me," and, of course, he received months of loving care from nurses. Camille gave up his search for love and transformed his life by dedicating himself to the love and care of others. As a result of the change of orientation from the needs of self to the welfare of others, he founded a clinic, got married, and became an excellent father. By loving others, he received the abundant love that had been absent during the first half of his life.

In his fiftieth Harvard reunion autobiography, Camille wrote:

> Before there were dysfunctional families, I came from one. My professional life hasn't been disappointing — far from it — but the truly gratifying unfolding has been into the person I've slowly become: comfortable, joyful, connected, and effective. Since it wasn't widely available then, I hadn't read the children's classic, *The Velveteen Rabbit*, which tells how connectedness is something we must let happen to us, and then we become solid and whole.
>
> As that tale recounts tenderly, only love can make us real. Denied this in boyhood for reasons I now understand, it took me years to tap substitute sources. What seems marvelous is how many there are and how restorative they prove. What durable and pliable creatures we are, and what a storehouse of goodwill lurks in the social fabric.

Godfrey Camille discovered in his journey through life that began with a bleak, painful childhood that no person is unlovable and that the most fundamental principle of human living is *a person is meant to love and to be loved*. Who can say deep down in his or her heart, "I do not want to love, nor do I want to be loved."?

19 The Best Moments of Human Life

When the habits of individualism are overcome through focused action or by accident, a person experiences what "connect" means in the defining characteristic of *Homo sapiens* — *the capacity to be connected to all that is.*

Each one of us has a non-stop monologue going on in his or her head. The inner monologue seems to be universal, although its content and intensity probably vary with culture. In an individual-centered culture, the inner monologue is an unending commentary on past and imagined events, invented arguments with others, and sweeping social and political commentaries, often mere unexamined opinions, with the facts ignored.

Traditional meditation practices, such as "watching" the breath and reciting a mantra, are spiritual techniques to stop the inner monologue.[378] In contrast to genuine dialogue, an inner monologue is unfocused, disconnected, and never arrives anywhere. Although an inner monologue is pointless, its tone reveals whether a person is chronically angry, complaining, self-pitying, righteous, or argumentative.

I suspect from my experience that an American alone in nature carries on an inner monologue that mindlessly jumps by association from thought to thought. My incessant, internal jabbering is absolutely stupid and is of no conceivable interest to anyone, not even to me. Like a monkey swinging from one tree to the next in a tropical forest, my mind leaps insanely from one topic to another. I even find it boring. Nevertheless, it is virtually impossible to shut it off — to be interiorly silent. Yet, without interior silence, I cannot really hear other persons, experience nature, or encounter the innermost depths of my being.

The Loss of Self

All of us have had rare occasions where our inner monologue is turned off for a moment. A friend of mine told me about her vivid memory of playing third base in the seventh grade forty years ago. She heard the crack of the bat, and a line drive headed toward her, traveling at least sixty miles an hour. To her, however, the ball moved so slowly that she watched the gradual rotation of the seams with fascination, and, of course, her gloved hand effortlessly caught the ball.

Another woman told me that she thought it impossible for any person to turn off his or her inner monologue. Then, one afternoon while walking in the woods with her husband, suddenly she was not preoccupied with the usual problems that cluttered up her mind. For a moment, she experienced the landscape, the wind, and the rhythm of walking. "It's wonderful!" she exclaimed to her husband, and the inner silence ended.

Several summers ago, I did some extensive mountain running in the White Mountains of New Hampshire. I remember running down the Tuckerman Ravine Trail, gliding from rock to rock at breakneck speed. One misstep and I could have broken an ankle or a leg. But that thought never entered my mind; I was without fear and totally confident. Entirely focused on what I was doing, the rocks stood out in bold relief. The varied shapes amazed me, and I marveled at the sea-foam green and bright yellow lichens on the rocks as they flew by me. I was so immersed in what I was doing that I was not a detached spectator, chattering away inside of my usual imaginary bubble. I was a person "lost" in union and in harmony with the rocks. On those rare occasions when the isolated self gives way to the loss of self, we become fully alive, and life needs no justification, for to be alive is joyful.

My experience of mountain running, of being totally absorbed in an activity, is called "in the zone" by athletes and "flow" by psychologists. Flow is the experience of performing an action at an optimal level, characterized by effortlessness and intense focus on the present; the self disappears, and the person becomes the performance. Rock climbers, surgeons, dancers, musicians, writers, chess players, mathematicians, indeed, actors in any field can lose the self, become the activity, and thereby experience flow — the loss of self.

Musician Barry Green observes that soloists, orchestral players, young students, and seasoned sessions players, alike, have experienced that "unique suspended moment when you actually become the emotional or sensory quality of the music — the colors, the water, the love."[379] An expert rock climber describes the same experience: "You are so involved in what you are doing [that] you aren't thinking of yourself as separate from the immediate activity. . . . You don't see yourself as separate from what you are doing."[380] A dancer says at times she becomes the dance: "Your concentration is very complete. Your mind isn't wandering, you are not thinking of something else; you are totally involved in what you are doing."[381]

The loss of self is not confined to specialized activities that require years to master. Any person can become lost in an everyday activity. A mother recounts the loss of self when she and her young daughter take turns reading to each other. "She reads to me, and I read to her, and that's a time when I sort of lose touch with the rest of the world, I'm totally absorbed in what I'm doing."[382] Many fiction readers enter into an imaginary world so completely that they become the characters, and the world around them recedes. Or even more simply, my friend mentioned earlier, while walking in the woods with her husband experienced the loss of self; she became one with the wind, rocks, and the trees when her inner monologue ceased momentarily.

With the loss of self in performance, our usual scattered attention disappears, and the mind becomes intensely focused, totally aware of the present. Such a concentration of attention Buddhists call "one-pointedness," meaning an interior state where all mental faculties are unified and directed toward one action, say sweeping the pebbles from a shrine, releasing an arrow from a bow, or sitting in meditation. For a tennis player, only the ball and the opponent exist; for a chess player, everything is excluded by the game's strategy.

When a person is completely in the present, not reflecting upon the past or worrying about the future, the senses are heightened: Vision is amazingly vivid, and hearing registers the subtlest changes in pitch and intensity. A violinist feels the tiniest movements of her fingers that produce palpable sounds. The music, her hands and fingers, her mind, all move in complete harmony. For rock climber Doug Robinson, "To climb with intense concentration is to shut out the world, which, when it reappears, will be as a fresh experience, strange and wonderful in its newness."[383] The great rock climber Yvon Chouinard describes how on the ascent up El Capitan's Muir Wall, he saw as if for the first time: "Each individual crystal in the granite stood out in bold relief. The varied shapes of the clouds never ceased to attract our attention. For the first time, we noticed tiny bugs that were all over the walls, so tiny that they were barely noticeable. While belaying, I stared at one for fifteen minutes, watching him move and admiring his brilliant red color."[384] Unlike our habitual looking without seeing, looking with real vision reveals the overwhelming beauty of mundane objects — clouds, snow, and granite.

Being in the present distorts a person's sense of time. If I am driving a car headed for a crash, I am not in the zone, yet I am suddenly in the present and totally aware of what is happening; time slows down so much that later I describe my experience as if it had happened in slow motion. A figure skater typically reports that a triple Axel that in clock time took several seconds to execute but lasted three times longer or more in subjective time. The opposite distortion of time also occurs. For a chess player or a novel reader, two hours may seem like two minutes.

For virtually every performer, the loss of self is desirable; personal problems vanish, the fear of failure disappears, mental clarity results, effortless performance happens, and joy ensues. The activity, then, becomes an end in itself. The desire for fame, public adulation, and even victory — in effect, all distractions from the marketplace and social life — disappear. The basketball great Bill Russell confesses that for him superb play took precedence over the desire for victory. Those perfect moments in basketball "were sweet when they came, and the hope that one would come was one of my strongest motivations for walking out there. Sometimes the feeling would last all the way to the end of the game, and when that happened, I never cared who won. . . . On the five or ten occasions when the game ended at that special level, I *literally* did not care who had won. If we lost, I'd still be as free and high as a sky hawk."[385]

The ancient Chinese sage Chuang Tzu taught that the desire to win a prize often ruins the skill of a contestant:

> When an archer is shooting for nothing,
> he has all his skill.
> If he shoots for a brass buckle,
> he is already nervous.
> If he shoots for a prize of gold,
> he goes blind or sees two targets —
> He is out of his mind!
> His skill has not changed. But the prize divides him.
> He cares. He thinks more of winning than of shooting —
> And the need to win drains him of power.[386]

Whether we recognize it or not, the loss of self in performance confirms our spiritual nature. When we become an object or an activity, comparative-religion scholar Toshihiko Izutsu avows, "Zen may be said to be already realized, whether one calls it Zen or not. Zen, however, requires that one should be in exactly the same state with regard to everything . . . One should become a bamboo. One should become a mountain. One should become the sound of a bell. That is what Zen means by the expression: 'seeing into the nature of things.'"[387] Eknath Easwaran, a disciple of Gandhi and the critically acclaimed translator of *The Bhagavad Gita*, *The Upanishads*, and *The Dhammapada*, concludes after years of meditation and study that "the goal of all spiritual seeking is to live in a state of self-forgetfulness permanently."[388]

Silence and self-forgetfulness are at the heart of Christian mysticism. Pseudo-Dionysius the Areopagite reports, apparently from his own experience, that "the higher we soar in contemplation the more limited become our expressions of that which is purely intelligible; even as now, when plunging into the Darkness that is above the intellect, we pass not merely into brevity of speech, but even into absolute silence of thoughts and of words."[389]

Not surprisingly, then, psychologist Mihaly Csikszentmihalyi and his coworkers have documented through extensive interviews that the best moments in life occur when "people become so involved in what they are doing that . . . they stop being aware of themselves as separate from the actions they are performing."[390]

In an informal survey I conducted, colleagues and acquaintances told me that the greatest joy in their lives occurred when they became lost in something outside themselves, in performing a Mozart piano sonata, in following the elegance of a mathematical demonstration, in seeing the beauty of the Canyonlands, or in helping a young child learn to walk. Joy is nature's way of telling us that we are fulfilling our nature. We find joy when we forget ourselves through knowing and loving those good things that are outside ourselves.

When we were children, the loss of self often came about because we were unself-conscious learners not afraid to fail. Later in school, we experienced failure as painful, for it brought bad grades from the teacher and sometimes laughter from fellow students. My first Spanish teacher, Señora Reverie de Escobedo, a wise woman, told me that her first goal in teaching a foreign language to adult learners was to undo the self-consciousness and fear of failure formed by years of prior schooling. Most of us have learned to see ourselves through the eyes of ever-critical onlookers, whether they are actually there or not. Before we act or speak, we first have to judge our actions or speech as we think others will. Some of us mentally rehearse everything we say, for our words have to be perfect; under such circumstances, we cannot act naturally or spontaneously.

After early childhood, we may have learned to play the piano so well that we could render a Mozart sonata flawlessly, yet not achieve an effortless, egoless performance. Optimal performance *does* require previous discipline and hard work, yet transcends habit and steps beyond the technically correct to the sublime.

Paradoxically, the universal course of our lives seems to move from aliveness and the loss of self to habit and deadness to aliveness and the loss of self. This strange trajectory begins with the naïve child free from social conditioning and ends with the sophisticated adult freed from mindless habit and an overriding desire for success. If we accede to the obvious truth that life is activity, then the highest human activity is the loss of self that results from making art, doing science, playing sports, educating the young, or caring for the old and disabled, from losing the self in an activity, from the death of the ego.

20 The Two Modes of the Mind

If we lack a word for an experience, we obviously cannot talk to others about it, and the experience, no matter how intense or unsettling, will fade from lack of understanding, and later we will be unable to judge whether it was significant or not.

Every scientist and mathematician, professional and student, whom I have known, can give countless occurrences in their own lives of hard, fruitless labor preceding effortless knowing. In the classroom or around the seminar table, I have heard an excited "I see it" innumerable times, but never in my life as a theoretical physicist did I reflect with colleagues about the meaning of such an experience. As a result, for years, I remained in the dark about the deepest aspects of what we are. My enlightenment came only after reading Plato and Aristotle. Through straightforward reflection about the interior life, accompanied by clear thinking, Plato and Aristotle discovered two aspects of the mind, which they called *nous* and *dianoia*.

(The Latin words *ratio* and *intellectus* correspond to the older Greek terms *dianoia* and *nous*. The Greek is used here because the Latin calls to mind the English words "reason" and "intellect," which often are used as synonyms. The English words are substantially more limited and less precise than either the Latin or the Greek.)

Labor

We commonly speak of knowing as defining, comparing, analyzing wholes into component parts, and drawing conclusions from first principles. Such discursive or step-by-step thinking the ancient Greek philosophers called *dianoia*. Sherlock Holmes is the epitome of *dianoia* at work. From the analysis of a cigar ash ground into a carpet, a scuff mark on a door, and the residue in a wineglass, he concludes the criminal is a lame aristocrat who resides in Kensington.

In discursive thinking, we either apply first principles or draw out their consequences. *Dianoia* operates step-by-step, almost in a mechanical fashion, repeating the same procedure again and again: A = B, B = C, therefore A = C; premise 1, premise 2, therefore, conclusion 1; etc.

What we frequently call thinking is not *dianoia* but the association of ideas. Consider the following scenario. In a seminar, the opening question is,

"What does Tocqueville mean by the equality of conditions?" Three minutes later, the participants are discussing the siege of Richmond. Through associations, the discussion hurried from the equality of conditions to political equality to slavery in the South to Ken Burn's *Civil War* to the Siege of Richmond. And soon, if the seminar leader did not intervene, the discussion would have moved on to the film *Gone with the Wind* and Clark Gable's last line: "Frankly, lady. I don't give a damn." There is, of course, no logical, physical, or psychological connection between the equality of conditions and Gable's exit line. If we step back and examine our "thinking," we would often find, perhaps to our amusement, that much of what we take for thinking is the mere association of ideas that has nothing to do with the actual structure of the world or the interior life.

Gift

Nous is the capacity for effortless knowing — to behold the truth the way the eye sees a landscape. The most famous story of sudden insight — of the "light bulb going off" — is the one told about Archimedes. The ruler Hiero II asked Archimedes to determine if the royal crown was truly made of pure gold or alloyed with silver. Archimedes knew that if the crown were not irregularly shaped, he could easily measure its volume and then check if its density was that of gold. But Archimedes could not figure out how to determine the volume of the crown. He was stumped; until one day, when he stepped into his bath, the solution suddenly appeared to him: A given weight of gold displaces less water than an equal weight of silver. He shouted, "Eureka! Eureka!" and ran naked through the streets of Syracuse.

Henri Poincaré, in his book *Science and Method*, describes how he discovered the relationship of Fuchsian functions to other branches of mathematics. After he had worked out the basic properties of Fuchsian functions, he left Caen, France, where he was living at the time, to take part in a geological conference at Coutances. The incidents of the journey made him forget his mathematical work. As a scheduled break from the conference, attendees went on a bus excursion. Poincaré reports, "Just as I put my foot on the step [of the bus], the idea came to me, though nothing in my former thoughts seemed to have prepared me for it, that the transformations I had used to define Fuchsian functions were identical with those of non-Euclidean geometry."[391]

When Poincaré returned home from the conference, he turned his attention to the study of certain "arithmetical questions without any great apparent result, and without suspecting that they could have the least connection with my previous researches. Disgusted at my want of success, I went away to spend a few days at the seaside, and thought of entirely different things. One day, as I was walking on the cliff, the idea came to me, again with the same characteristics of *conciseness, suddenness*, and *immediate certainty*, that arithmetical transformations of indefinite ternary quadratic forms are identical with those of non-Euclidean geometry."[392]

Mathematician Carl Friedrich Gauss tried unsuccessfully for two years to prove an arithmetical theorem. In a letter to a colleague, he revealed, "Finally, two days ago, I succeeded, not on account of my painful efforts, but by the grace of God. Like a sudden flash of lightning, the riddle happened to be solved. I myself cannot say what was the conducting thread which connected what I previously knew with what made my success possible."[393]

Astrophysicist Fred Hoyle relates how the solution to what seemed an intractable mathematical problem in quantum physics came to him while driving a car on the road over Bowes Moor, much as "the revelation occurred to Paul on the Road to Damascus: My awareness of the mathematics clarified, not a little, not even a lot, but as if a huge brilliant light had suddenly been switched on."[394]

What Hoyle describes as a "huge brilliant light" and Poincaré calls a "sudden illumination"[395] and Gauss terms "a flash of lightning," the ancient Greek philosophers named *nous*, whose characteristics are "conciseness, suddenness, and immediate certainty," as Poincaré noted. We, clearly, cannot command *nous*; the direct grasp of a truth is a gift, not something we can turn on or off at will, for if we could, we would.

Sometimes the direct grasp of a truth is astonishing. Extraordinary mathematicians such as Gauss and Riemann wrote down theorems without having any idea of how to prove them. Other mathematicians struggled to demonstrate those theorems and eventually did, although the proofs required pages and pages of complex reasoning. The Indian mathematician Srinivasa Ramanujan because of his limited education had no clear idea of what it means to prove a theorem. Yet, Ramanujan directly grasped many remarkable mathematical relations. The English mathematician Godfrey H. Hardy was astonished by the theorems that Ramanujan sent him: "I have never seen anything in the least like them before. A single look at them is enough to show they only could be written down by a mathematician of the highest class."[396] Some of Ramanujan's theorems Hardy proved with great difficulty, others eluded him, although he was convinced they were true, for "no one would have the imagination to invent them."

The direct grasp of a truth is generally accompanied by joy, especially when preceded by intense, fruitless discursive thinking, as the following example shows. Astronomer Maarten Schmidt describes his discovery of quasars as a "once-in-a-lifetime experience . . . the biggest in astronomy, I suppose, since the discovery that there exist galaxies besides our own. I remember the day, the hour, it happened. It was early in the afternoon of February 5, 1963. Until then, for months and months, I had been puzzling over this inexplicable thing, and then I had this sudden perception, and within half an hour everything was clear to me."[397]

Overjoyed, Schmidt desired to share the insight he received. He called his colleague Mitchell Wilson into his office. Wilson remembers how Schmidt "showed me his photographic plate of the spectrum of the stellar object co-incident with the radio source called 3C-273, which I had also looked at, and then he gave me this remarkable explanation of it! He made it so obvious I couldn't bear it! I screamed! Literally, I shouted! I was that excited! It was marvelous!"[398]

After receiving a sudden insight, every physicist and mathematician I have known experiences two impulses, one from the interior life and the other from the marketplace. The first impulse is to rush out and show what he or she has received to anyone who will listen. What motivates the recipient is an experience of a beauty or a truth that is so elegant or so profound that no person should miss out on having it. The other impulse is immediately to write a paper to ensure that he or she receives recognition for "his" or "her" discovery, not someone else.

The interior impulse to share beauty and truth is universal. Many years ago, my children and I hiked up Pat's Peak in New Hampshire for the first time. It was one of those beautiful fall days in New England that you read about. The air was cool and crisp, the sky cloudless, and the foliage ablaze with color. On our way up the mountain, we met a stranger coming down the trail. He excitedly told us about a magnificent view on the other side of the mountain and gave us detailed instructions, so we would not miss finding the special spot he discovered. A complete stranger doing us a good turn! He did not ask anything in return; he gave to us freely. Truth and beauty draw human beings together and impel them to share the interior life.

Our knowing, generally, entails both *dianoia* and *nous*. We know directly that "the whole is greater than the part" and that "every human being by nature desires to know." The child knows "triangle" as well as any mathematician; yet, it will take years of study and laborious toil before he or she will understand algebraic topology. *Nous* immediately grasps the whole of the argument in a flash. When we comprehend how all the parts fit together and why they are there, we spontaneously shout, "I see it." The phrase "I see it" accurately describes the experience of effortlessly grasping the whole of an argument or the connection between seemingly disparate things, for *nous*, as we have seen, is like the physical eye gazing upon a landscape.

To know in the deepest sense is not to possess accurate information, say the names of the anatomical parts of a rabbit, or to have a method that turns out correct answers, say a mass spectroscope that determines molecular masses. To know profoundly is to experience an insight, a revelation of a deep and significant truth. Such knowing cannot be summoned by the will, although insight is usually preceded by hard work and the clearing away of major errors and wishful expectations. Sometimes, insight occurs after only giving up. In contrast to the truths arrived at through hard labor, intuitive insights are gifts. Consequently, *dianoia* usually results in pride, *nous* almost always in gratitude.

Dianoia and *nous* are not confined to mathematics and science but are fully present in any intellectual activity. Mozart, in a letter, describes how he composes music. He informs his correspondent that musical composition involves two elements: The musical parts are put together according to the rules of composition (*dianoia*); and the joy the composer receives when he sees the composition as a whole, much as the eye sees a beautiful landscape (*nous*): "When I feel well and in good humor, or when I am taking a drive or walking after a good meal, or in a night when I cannot sleep, thoughts crowd into my mind as easily as you could wish. Where do they come from? I do not know, and I have nothing to do with it. Those which please me I keep in my head and hum them; at least others have told me that I do so. Once I have my theme, another melody comes linking itself with the first one in accordance with the needs of the composition as a whole; the counterpoint, the parts for each instrument and all the melodic fragments at last produce the complete work. Then my soul is on fire with inspiration. The work grows; I keep expanding it, conceiving more and more clearly until I have the entire composition finished in my head although it may be long. Then my mind seizes it, as a glance of my eye would a beautiful picture or a handsome youth. It does not come to me successively, with various parts worked out in detail, as they will later on, but it is in its entirety that my imagination lets me hear it."[399]

Those poets in Modernity who explore the depths of the interior life acknowledge publicly that all artistic creation rests upon gifts. The Polish poet Czeslaw Milosz, in the autobiographical essay "Catholic Education," confessed, "I felt very strongly that nothing depended on my will, that everything I might accomplish in life would not be won by my own efforts but given as a gift."[400]

Theodore Roethke in "On 'Identity'," a lecture delivered at Northwestern University, told how at the age of forty-four, he was in a particular hell for a poet, a longish dry period; he thought he was finished as a poet. For weeks at the University of Washington, he had been teaching five-beat line to aspiring poets but felt a total fraud because he could write nothing himself. "Suddenly, in the early evening, the poem 'The Dance'[401] started, and finished itself in a very short time — say thirty minutes, maybe in the greater part of an hour, it was all done. I felt, I knew, I had hit it. I walked around, and I wept; and I knelt down — I always do after I've written what I know is a good piece." The gift arrived, and Roethke thanked the source and "wept for joy."[402]

We must not be misled into thinking that *nous* only operates in geniuses, giving them direct grasp of profound truths. When a student of Euclid's

Elements truly grasps a demonstration, he or she shouts with joy, "I see it." Through *nous*, the student, in a flash, immediately grasped the whole of the argument. Without *nous*, any rational argument cannot be apprehended. In other tutorials, scientific, poetic, and philosophic insights occur. A student may suddenly grasp how the *Iliad* forms an integral whole or why Socrates in the *Phaedo*, on the day of his execution, spins philosophical "tales" about the immortality of the soul to comfort his distraught friends.

The above discussion of *dianoia* and *nous* rests on straightforward observations of the interior life and consequently should not be controversial. As predictable, the ancient and modern ways of understanding the two modes of the mind are opposed. For Plato and Aristotle, *dianoia* is human and *nous* divine.

Plato rightly contends that to directly grasp a truth is superior to arguing to it. Consequently, *dianoia* is always in the service of *nous*. In the *Seventh Letter*, he explains how "after practicing detailed comparisons of names and definitions and visual and other sense perceptions, after scrutinizing them in benevolent disputation by the use of question and answer without jealousy, at last in a flash of understanding . . . the mind, as it exerts all its powers to the limit of human capacity, is flooded with light."[403]

In Modernity, *nous* is denied, ignored, or misunderstood. As a result, the origin of all knowledge is taken as unaided human effort. Almost without exception, scientists and mathematicians claim to be the sole source of their discoveries. When intellectual insight is acknowledged as unwilled, it is attributed to the hidden workings of an unconscious mind, not to a super-natural gift. Nothing in science or mathematics is seen as a gratuitous gift from Divine Mind, or God. Scientists and mathematicians, with their denial of *nous*, proclaim they are beholden to nothing beyond themselves. Instead of being thankful, they are prideful, for they believe that all knowledge and invention results from human labor alone. How different was Pythagoras! After the remarkable theorem that now bears his name was revealed to him, Pythagoras in gratitude sacrificed an ox to the gods.

In mid-life, I had left materialism behind as a pernicious myth of the Dark Ages of Science. To call *nous* a product of unconscious thought as Poincaré and other scientists had was to beat the same old drum — the brain alone explains all interior life, although now we know with certainty that neuronal activity by itself cannot account for even perception. It made more sense to me to understand *nous* as the divine element within each person than to attribute our highest intellectual life to the murky unconscious mind hidden somewhere in the brain. To me — and I was in the excellent company of Plato, Aristotle, Augustine, and Aquinas — *dianoia* was strictly human, and *nous* was the divine element within us.

In search of Divine Mind, I pressed on to re-evaluate the updated, classical "proofs" for the existence of God.

21 Atheism v. Theism

Since I, like you, am living in the Kali Yuga, the Dark Age of Hindu mythology, when all the great faiths are on the wane, I was not surprised by a recent survey of religious belief among the leading scientists of the United States as represented by the members of the National Academy of Sciences that found a "near universal rejection of the transcendent," with the belief in a personal God and in human immortality a mere seven percent.[404] Virtually every beginning physics student has a killer argument for atheism and the non-existence of the transcendent; I know I did. With my two Catholic fellow students, I argued as follows: 1) only matter and void exist; 2) God is a nonmaterial being; and 3) therefore, God does not exist. Years later, I realized that the collapse of materialism made this killer argument invalid, and for me, the existence of God became an open question. Of course, the reality of the nonmaterial does not mean that God, the supernatural, or the transcendent exists. Schrödinger and Weizsäcker convinced me that visual perceptions are nonmaterial, but Oliver Sacks' patient, the one who suffered a thrombosis in the posterior circulation of the brain, clearly exhibited that with the immediate death of the visual parts of brain, a lifetime of visual experience disappears and that the immaterial may not exist without matter.

To see if biologists, astronomers, and physicists disbelieve in the existence of God for reasons other than materialism, I picked up a copy of *The Portable Atheist: Essential Readings for the Nonbeliever* with the expectation of finding many killer arguments against the existence of a deity, for surely scientists who pride themselves on rational evidence must have produced scores of justifications for their belief.

What a surprise! I found emotional polemics against the ill effects of religion, an irrelevant and mundane observation given the dismal record of human-run institutions; men and women are finite and almost always act from a narrow cultural perspective. To me, the essays that contended that human suffering and death mean a compassionate God does not exist seemed to be a quarrel with God about the world He created, not a reasoned denial of His existence. Only Richard Dawkins, an evangelical atheist and the author of *The God Delusion*, claimed an "unanswerable" argument in the chapter "Why There Almost Certainly Is No God," an argument that stumped every theologian who encountered it, or so he said.[405]

Dawkins calls his argument the Ultimate Boeing 747 Gambit. The name comes from a metaphor of Fred Hoyle, the same maverick astrophysicist who coined the term "Big Bang." Hoyle claimed life on Earth must have originated some other place because the probability of life arising from a chemical soup on primordial Earth is vanishingly small, smaller than the probability of a tornado sweeping through a junkyard resulting in a working Boeing 747 airliner.[406]

Although Dawkins spreads his argument out over ten pages, it can be presented in two paragraphs. He observes that evolutionists, like himself, and creationists agree that complex things like plants and animals came about by chance alone is improbable. However, unlike creationists, Dawkins sees that natural selection provides an elegant solution to this problem of improbability. Each small step in natural selection is improbable, but not prohibitively so. The accumulation of a large number of these small, slightly improbable steps adds up to a very, very improbable result, an outcome unfathomable to the non-scientific mind. Dawkins jeers that doofuses like creationists just don't get it. They rightly believe that plants and animals did not come about by chance but wrongly conclude that organisms must therefore have been designed by God.

Dawkins asserts that any God capable of designing a universe must be supremely complex. Just as the blacksmith is more complex than the horseshoe he fashioned, God is more complex than his creation. But complex things are improbable. Therefore, God is more improbable than that the universe came about by chance alone.

At a loss by what he meant by "complex," I searched about and discovered that using natural selection as a guide, Dawkins in his book *The Blind Watchmaker* defines a complex object as "something whose parts are arranged in a way that is unlikely to have arisen by chance;"[407] this definition is implicitly assumed in his "unanswerable" argument for atheism. Therefore, hidden in Dawkins' killer argument is that God has parts, that is, he is a material being. The legitimate conclusion Dawkins should have drawn is that God cannot have a body, a trivial result with which every theist would concur. In the thirteenth century, St. Thomas Aquinas formally presented the case that "God is nowise composite, but is altogether simple," meaning he is without quantitative parts, nor composed of matter and form.[408]

The other error in the "unanswerable" argument is that complex things are improbable. Three possibilities exist for the production of complex things: 1) chance alone (highly improbable), 2) natural selection (very improbable, but possible, as Dawkins correctly argued), and 3) mind (very probable in certain circumstances). An example of the third possibility is last week I went hiking near Santa Fe, New Mexico, my hometown, and in the wilds came across an inch-square piece of hard clay with black, narrow stripes. I knew the probability of such a piece being formed by geological activity is vanishingly small; however, if the piece were a pottery shard, then it was certainly the result of a human mind. Any place where humans have been present, I and every other scientist expect to find complex objects not formed by nature. The probability of finding artifacts where ancient peoples once lived is a near certainty. If, indeed, the universe were created by a Divine Mind, then the probability of finding somewhere in it entities more complex than desert sand and crystals would be a certainty. (Here Divine Mind is at best a metaphor, a human way of speaking, see Chapter 23.)

Dawkins with his Boeing 747 Gambit avoids the central question of whether in some mysterious way Divine Mind could be behind the complex order in nature. From Dawkins' failed attempt, I concluded there is no killer argument for atheism, and that left open the possibility of a conviction-carrying argument for theism.

I decided to test the three killer arguments of medieval theology against modern science, for I, a seeker trained in theoretical physics, would not be persuaded by arguments that rely on an Aristotelian understanding of nature.

The ontological argument came to St. Anselm in 1078. In the third person, he describes the intellectual ecstasy the discovery gave him. "Behold, one night during Matins, the grace of God shone in his heart and the matter became clear to his understanding, filling his whole heart with immense joy and jubilation."[409]

Eight hundred years later, Bertrand Russell, as a young philosopher, experienced the same enthusiastic, although brief, acceptance of the ontological argument. In his essay "My Mental Development," Russell recorded his epiphany: "I remember the precise moment, one day in 1894, as I was walking along Trinity Lane, when I saw in a flash (or I thought I saw) that the ontological argument is valid. I had gone out to buy a tin of tobacco; on my way back, I suddenly threw it up in the air, and exclaimed as I caught it: 'Great Scott, the ontological argument is sound.'"[410]

The Ontological Argument in Modern Form

1. By definition, God is the greatest possible being that can be conceived.
2. God exists as an idea in the mind.
3. A being that exists as an idea in the mind *and* in reality is greater than a being that exists only as an idea in the mind.
4. Thus, if God exists only as an idea in the mind, then we can conceive something that is greater than God.
5. But that we can conceive a being greater than the greatest possible being that can be conceived is a contradiction.
6. Therefore, God exists in reality, not just as an idea.

When I first heard the ontological argument, I was immensely disappointed by this supposedly killer argument for the existence of God. My first thought was the argument had to be wrong, for how can the existence of anything follow from a definition. Always in mathematics when a definition is given, an example must be shown; otherwise, an elaborate theory may be constructed about nothing. Consider a married bachelor. He lives by himself, and yet everyday he is nagged by his wife.

Still, I could not find a flaw in the logic of the ontological argument. Later in life, Russell confessed, "It is easier to feel convinced that the ontological argument must be fallacious than to find out precisely where the fallacy lies."[411] A philosopher friend of mine pointed out the difficulty with this abstract proof — what philosophers call an *a priori* demonstration — of God's existence. He told me that Russell had refined Immanuel Kant's refutation of the ontological argument. "Being" or "existence" is not a property that can be attributed to objects. I said, "huh," and he went on to explain that existence is what an object must have to have a property. For example, in the sentence "the table is clean," the table must exist to have the property clean. Existence is not a real property like clean and cannot be predicated of any object. In the sentence, "the table *is*" the predication *is* adds nothing. Formally, to predicate the property Y of X means there exists some X, such that X has the property Y.

My friend told me that Kurt Gődel, perhaps the greatest logician ever and deified by ordinary philosophers, constructed a contemporary version of the ontological argument using modal logic, whatever that is. But no ordinary mortal, myself included, would be persuaded God exists from an arrangement of symbols on a page. Tragically, Kurt Gődel starved himself to death; he had convinced himself that someone was trying to poison him, so he refused all food. The attempt to grasp the world, or even one's self, by logic alone without experience or fundamental principles of nature leads to madness.

The failure of the ontological argument is a victory for the atheists.

The two other killer arguments for God rely on observations of nature. They are the cosmological argument and the argument from design.

The Cosmological Argument in Modern Form

The cosmological argument may appear abstract and contrived, but it is rooted in a universal experience. Every child asks, "Where did I come from?", and receives the answer from mommy and daddy. Further questioning from the child leads to a long chain of ancestors. In virtually every culture, the chain of ancestors terminates with the first parents of humanity. For instance, in Judaism, Christianity, and Islam, humankind begins with Adam and Eve, in Australian Aboriginal mythology with Wurugag and Waramurungundi, and in Norse folklore with Ask and Embla.

The cosmological argument is a philosophical story about the origin of the cosmos. Nothing in the world is the cause of its own being. Everything that comes into existence has a cause outside of itself for its existence. But every cause in the world has a cause, and that cause has a cause, and so on. To avoid an infinite number of causes, every chain of causes must eventually terminate with the First Cause.

Until the twentieth century, science could not tell a story about the origin of the universe. Newtonian physics, when applied to the universe as a whole, led to contradictions and paradoxes. The Newtonian Cosmos must be infinite with the matter perfectly balanced; otherwise, the universe under gravitational attraction would collapse. Such a universe is inherently unstable; the slightest perturbation would cause an implosion. German astronomer Heinrich Olbers, in 1823, correctly argued that an infinite, static universe contradicts the most basic observation of astronomy — the night sky is black. In the Newtonian Cosmos, everywhere the eye looks, it would see a star, and thus the night sky would be bright, in fact, infinitely bright.

In the twentieth century, Einstein's theory of general relativity stepped beyond Newtonian physics to integrate gravity, space, and time. For the first time, physicists had a theory to investigate in detail the structure, the origin, and the destiny of the whole universe. Cosmologists now know that our universe is the aftermath of a gigantic expansion of matter. Present size and rate of expansion indicate that the universe began about 13.8 billion years ago. At the very beginning of the universe, in the quantum era, the entire universe was roughly a million billion billion billion times smaller than a proton. The density at that point staggers the imagination. Imagine all the matter and energy of what would become planets, stars, and whole galaxies contained in a space next to nothing in size! At time zero the density was infinite with no extension in space at all, what physicists call a singularity that marked the beginning of matter, time, and space. (See Figure 21.1.)

The scientific discovery of the universe's birth posed to cosmologists what physicist Steven Weinberg called the "problem of Genesis."[412] Our universe — including matter, energy, space, and time — is almost certainly a one-time event that had a definite beginning. The physical universe is not eternal. But something must have always existed, for if ever absolutely nothing existed, then nothing would exist now, since nothing comes from nothing. The material universe cannot be the thing that always existed because matter had a beginning some thirteen billion years ago. Therefore, whatever has always existed is nonmaterial. The only nonmaterial reality seems to be mind. If mind is what has always existed, then matter must have been brought into existence by a mind that always was, by an intelligent, eternal being who created all things *ex nihilo*. Such a being is what theologians mean by God.

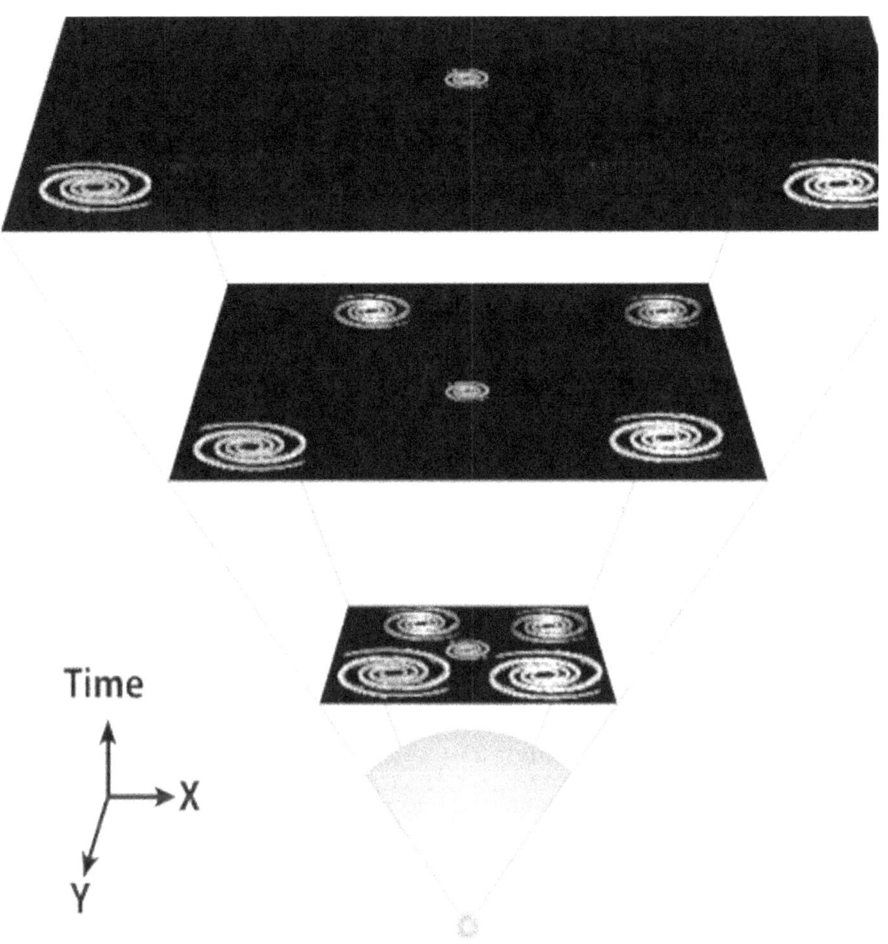

Figure 21.1. The expansion of a two-dimensional universe after a Big Bang (Singularity). (Milagli, Can StockPhoto.)

From the standpoint of science, the most sophisticated way to avoid the theological consequences of the Big Bang is to invoke quantum physics, where the everyday rules of cause and effect do not apply. Polonium-210, supposedly used by former KGB agents to poison enemies of the Russian state, is a very rare, unstable element. Given a sample of Polonium-210, after 138 days, only half remains. Polonium-210 nuclei decay spontaneously; quantum physicists, in the 1920s, discovered that no cause exists for why a particular nucleus decays at a particular time. All nuclear decay is governed by chance; only a probability can be assigned to the decay of a particular nucleus.

Quantum effects must have been once important on a cosmic scale because shortly after the Big Bang, the universe was much smaller than a proton. Accordingly, the universe may have just happened, just popped into existence, the way that elementary particles are now thought to pop into existence and then disappear. Thus, there is no cause for the universe, either natural or supernatural.

This rebuttal of the cosmological argument does not end conclusively. Opponents point out that whether quantum physics, a theory of atoms and elementary particles, can be applied to the universe as a whole is problematical because such fundamental concepts as probability and the wave function are then meaningless. The modern cosmological argument does not terminate with a victory for either the atheists or the theists.

The Argument from Design

In both manmade and natural objects, design is apparent. Matter by itself cannot produce microchips, digital cameras, or Boeing 747s; human minds are needed. Analogously, matter on its own cannot bring into being petunias or Siberian tigers; a mind is needed, and the only candidate is a Divine Mind.

Natural selection, supposedly, crushed the argument from design, first given by William Paley, in 1802. Darwin showed unequivocally that a small series of improbable, small blind steps *could* result in apparent design without the guidance of a designer. We should note that the fundamentalist revolt against teaching Darwinian evolution in the Bible Belt and the invention of "creation science" may be a historical aberration connected to a brand of Protestantism wedded to a literal interpretation of the Bible. Oddly, Augustine in a lengthy work entitled *The Literal Meaning of Genesis* (don't let the title fool you) argued that in the beginning God created all living things not immediately as actual individuals, but "invisibly, potentially, in their causes, as things that will be in the future;" that is, the fullness of nature would unfold in time, a fifth-century, theological statement of evolution.[413]

The most recent argument for design comes from physics, not biology. Abundant evidence from the physical sciences reveals that our universe is bio-friendly. For instance, if the strong nuclear force that binds protons and neutrons together in the nuclei of atoms were slightly stronger, the diproton would exist, making ordinary hydrogen catastrophically explosive. Hydrogen would be rare, and stars like the Sun that live a long time by burning hydrogen could not exist. If the strong force were half its present strength, hydrogen would not burn at all, and there would be no heavy elements. If, as seems likely, life requires a star like the Sun, supplying energy at a constant rate for billions of years, then for life to be possible, the strength of the strong force must be within a narrow range. If the strong force were slightly stronger, no hydrogen; if less than half its present strength, no heavy elements. Either way, life is impossible.

The same pattern recurs with many other fundamental constants of nature. Exploding stars (supernovas) play an essential part in the chemical evolution of galaxies. Right after the Big Bang, the material of the universe was primarily hydrogen and helium. Heavier elements were later synthesized inside of stars. These stars aged, eventually exploded, and dispersed into space the heavier elements in their interiors. The carbon in the sugars we eat, the iron in our red blood cells, and the calcium in our bones came from supernovas. But exploding stars would not be possible if the weak interaction that governs the decay of neutrons into protons, electrons, and neutrinos was slightly altered. If the weak interaction were weaker, neutrinos could not exert enough pressure on the outer envelope of a star to cause a supernova explosion. If the weak interaction were stronger, the neutrinos would be trapped inside stars and no supernovas would occur. Either way, the heavier elements manufactured in the core of stars could not be injected into a galaxy to later form second-generation stars like our sun and planets capable of supporting life. Hence, the weak interaction appears suited to making life possible in the universe.

Hoyle discovered a nuclear resonance that makes possible the synthesis of carbon nuclei in stars. Two helium nuclei join together to form the unstable nucleus of beryllium 8, which sometimes before fissioning, absorbs another helium nucleus, forming carbon 12 in an excited state. If the energy of this excited state of carbon 12 were just slightly higher, almost all the beryllium 8 nuclei would fission into helium nuclei before carbon could be formed and nuclei heavier than boron would not exist. The universe, then, would consist almost entirely of hydrogen and helium — and life would be impossible. Years later, Hoyle concluded that the precise details of nuclear physics that resulted in a bio-friendly universe were a "put-up job." Abandoning his earlier and vehement anti-God stance, he said, "A common sense interpretation of the facts suggests that a superintellect has monkeyed with physics, as well as with chemistry and biology, and that there are no blind forces worth speaking about in nature."[414]

The list of strangely fortunate physical properties that make life possible goes on and on. Without the "put-up job" in atomic physics, water would not exist as a liquid, chains of carbon atoms would not form complex organic molecules, and hydrogen atoms would not form breakable bridges between molecules.

The most striking "put-up job" is the fine-tuning of the expansion of the universe. If the universe were expanding too slowly, it would re-collapse into oblivion. If the universe were expanding too fast, matter would become isolated and galaxies would not form. In reality, the expansion of the universe is exquisitely tuned to ensure the formation of galaxies, stars, and planets.

The discovery, in the 1990s, that the expansion of the universe is accelerating revealed how incredible this fine-tuning is. Previously, Einstein added the cosmological constant to the field equations of general relativity to ensure the universe is stationary. Shortly after Hubble discovered the universe is expanding, Einstein called the introduction of the cosmological constant the "biggest blunder of my career," and to account for an expanding universe, he set the cosmological constant equal to zero. Today, with the expansion of the universe accelerating, the cosmological constant must be nonzero, but small. Weinberg pointed out that a nonzero cosmological constant could result from quantum fluctuations of the vacuum.[415] He calculated that particle physics gives a cosmological constant 10^{120} larger than the observed value, which has been called "the worst theoretical prediction in the history of physics!"[416] If the cosmological constant were slightly larger than what it actually is, then the early universe would have expanded so rapidly that stars and galaxies would not have formed and life would never have emerged. If the cosmological constant were slightly smaller, then the early universe would have contracted so rapidly that the simplest atoms could not form. The obvious explanation for the observed value of the cosmological constant is the universe is to an astonishing degree designed for life.

The evidence that the universe is uniquely suited for life is undeniable; no physicist doubts this. The strong anthropic principle states that the universe must have those properties that allow intelligent life to appear at some stage in its history. After the Big Bang, the universe appears to be aiming at life and at intelligent observers. Physicist John Archibald Wheeler asks, "What possible sense it could make to speak of 'the universe' unless there was someone around to be aware of it." As we saw in Chapter 2, to construct a map of the universe *as if the scientist were not part of it* is a slight-of-hand trick, a stratagem that gives determinism the appearance of being true. Wheeler argues that "awareness demands life. Life in turn, however anyone has imagined it, demands heavy elements. To produce heavy elements out of the primordial hydrogen requires thermonuclear combustion. Thermonuclear combustion in turn needs several times 10^9 years cooking time in the interior of a star. But for the universe to provide several times 10^9 years of time, according to general relativity, it must have a reach in space of the order of several times 10^9 light years. Why then is the universe as big as it is? Because we are here!"[417]

A startling reversal of perspective; in the seventeenth century, Pascal lamented, "The eternal silence of these infinite spaces fills me with dread;"[418] for we moderns, the diameter of the observable universe is approximately a trillion trillion miles; the mind-bogglingly size of the universe crushes us into insignificance. When the vastness of the universe is seen as making life possible, we are elevated in the scheme of things. Because we are here, the universe must be vast.

A universe aiming at the production of intelligent observers implies a mind directing it; for matter on its own cannot aim at anything. A mind that directs the whole universe, all the laws of nature, and all the properties of matter to a goal is called God by theologians.

Most physicists hate the anthropic principle because of the obvious implication that God exists. The latest way to accept the evidence that the universe is uniquely suited for life *and* deny the existence of God is to appeal to a multiverse, an appeal that can be rooted in either cosmology or particle physics. Some cosmologists adhere to the theory of eternal inflation, where our Big Bang is just one of an infinite number that has taken place in an infinite universe. According to string theorists, our universe is one of billions and billions of universes — 10^{500} to be more or less exact.[419] The number 10^{500} is mind-bogglingly large. The number of grains of sand that could be densely packed into the visible universe is infinitesimal compare to 10^{500}.

In the 10^{500} universes of string theory or the infinite number of universes of eternal inflation, physical properties are assumed to occur by chance so that a universe like ours "fine-tuned" for life is bound to appear. In this way, chance replaces God. Furthermore, string theorists and cosmologists agree that we are trapped forever in our own Big Bang, and presumably, no observational evidence can ever confirm the existence of other universes.

The adoption of the multiverse strategy in order to avoid the Hand of God in our universe forces theoretical physicists to give up their dream of explaining all the properties of our universe in terms of a few fundamental natural laws and several physical constants. Before the multiverse, particle physicists hoped to derive all the elementary particle masses in terms of one or two fundamental masses and to understand the four fundamental forces of nature in terms of a single force. But in the multiverse strategy to avoid the existence of God, each universe has its own physical constants that occurred by chance and thus can never be derived from any theory. Physics, then, is an environmental science that depends upon the particular universe scientific observers happen to inhabit. "If this is true," a dejected Weinberg concludes, "then the hope of finding a rational explanation for the precise values of quark masses and other constants of the standard model that we observe in our big bang is doomed, for their values would be an accident of the particular part of the multiverse in which we live."[420]

In addition, some physicists maintain the laws of physics are merely local bylaws that vary from universe to universe, for there is no reason to assume otherwise. Cosmologist Lawrence Krauss argues that the idea of eternal, elegant laws that rule nature, the laws sought by Copernicus, Galileo, Newton, and all subsequent physicists, rests upon the Creator-God of Judeo-Christianity, Who commands nature through nonmaterial laws. For Krauss the choice is stark; either the laws of physics are determined by a Creator-God or result from random, physical processes. For scientists to rid themselves once and for all of a Creator-God, they must accept that physical laws "could be almost anything,"[421] and that no fundamental theory of matter exists; each universe in the multiverse has its own physical laws, although quantum fluctuations probably occur in each universe. With this view, at the most fundamental level, every universe is irrational, outside the domain of any future science. The desire, then, to explain our universe's features in terms of fundamental causes and principles must be repressed.

In summary, the multiverse strategy to avoid the existence of God requires the suspension of scientific argument, demands a belief in the existence of at least a nearly infinite number of unobservable universes, and makes hopeless the desire to understand the most fundamental aspects of our physical universe in terms of experiment, mathematics, and logic.

To a theist, an atheist who believes in a multiverse will believe anything, even in an infinite number of invisible objects with 50 billion galaxies or more. In the past, physicists made fun of medieval theologians arguing about how many angels could dance on the head of a pin. Now, these modern devotees of reason invoke with a straight face billions and billions and billions of unobservable universes to explain the astounding properties of one — ours — a clear violation of Ockham's razor, the intellectual restrain that entities should not be multiplied beyond necessity. Although dressed up in scientific language, the multiverse theory requires a greater leap of faith than a belief in God.

Surprisingly, the underpinnings of the multiverse — eternal inflation and string theory — lack experimental verification. Physicist Paul Steinhardt co-authored seminal papers on inflationary cosmology, yet later argued that if eternal inflation were true, then "everything that can physically happen does happen an infinite number of times. No experiment can rule out a theory that allows for all possible outcomes. Hence, the paradigm of inflation is unfalsifiable."[422]

Also, the implications of an infinite number of Big Bangs in a multiverse governed by deterministic laws borders on the absurd. In such a multiverse, every finite system with a finite number of states occurs an infinite number of times. Thus, innumerable planets like Earth occur. On countless Earths, the Nazis lost World War II, and on the same number of planets, they won.

In a 1987 interview, Feynman pointed to a fundamental problem with the work of string theorists: "I don't like that they're not calculating anything. I don't like that they don't check their ideas. I don't like that for anything that disagrees with an experiment, they cook up an explanation — a fix-up to say 'Well, it still might be true.'"[423] For more than twenty-five years after Feynman's criticisms, at least 3,500 high-energy theorists have struggled with string theory, creating elegant mathematics and a better understanding of quantum field theory, but not one of these theorists came up with an experimental test of string theory. No wonder Sheldon Glashow, one of the principal architects of the standard model of particle physics, asks ironically, "Are string thoughts more appropriate to departments of mathematics, or even to schools of divinity, than to physics departments?"[424]

Biologist Richard Lewontin makes explicit the underlying faith that leads scientists to accept a universe that is at heart irrational and thus unknowable rather than to acknowledge that the Hand of God shaped nature: "...we have a prior commitment, a commitment to materialism. It is *not* that the methods and institutions of science somehow compel us to accept a material explanation of the phenomenal world, but, on the contrary, that we are forced by our *a priori* adherence to material causes ... Moreover, that materialism is absolute, for we cannot allow a Divine Foot in the door."[425]

Lewontin and most scientists are true believers in materialism, possessing an absolute faith that matter and its workings will eventually explain everything in the universe. But such faith has already failed at the most basic level; brain function alone cannot account for the simple experience of seeing, hearing, tasting, or smelling. All human beings, scientists and laypersons, live in the nonmaterial world of the smell of lavender, the deep resonance of a cello, the beauty of a sunset over an ocean, the wonder evoked by the night sky, the elegance of Euclid's demonstration of the infinitude of prime numbers, the very world that materialism cannot explain. If only matter existed, then we would have no interior life; we would be mindless things like rocks and volcanoes.

Like every ideology — political, religious, or intellectual — materialism at some point is closed to reason so that eminently intelligent adherents must make exceedingly idiotic pronouncements: "All of us human beings and all the objects with which we deal are essentially bundles of simple quarks and electrons;"[426] "You're nothing but a pack of neurons;"[427] and "You're a gigantic lumbering robot manipulated by genes."[428]

For me, materialism is far from absolute because it has failed again and again. No one can doubt we live in a bio-friendly universe. The multiverse strategy to avoid the existence of God borders on the absurd and furthermore declares the core of the physical world is irrational and thus unknowable. Giving up the ideology of my youth, abandoning personal preferences formed by culture, and following reason wherever it leads, I had no choice but to "allow a Divine Foot in the door."

22 Physics, Beauty, and Divine Mind

Last week, my wife, a painter-friend of ours, who wishes to be anonymous, and I did the Friday night walk down Canyon Road, the site of numerous galleries in Santa Fe, New Mexico, a small town that is the third-largest art market in the United States. Halfway down Canyon Road, we stopped in at a contemporary gallery that had a new show featuring conceptual art that our painter-friend was dying to see. Inside, I gazed at one wall of the gallery painted red with white block letters pronouncing BELIEF + DOUBT = SANITY; on another wall, I encountered a photograph of a human brain with the caption A THINKING MACHINE; next to the photograph was a large white canvas with the black letters PURE BEAUTY. Puzzled by what I saw, I mumbled, "What pointless crap. What happened to beauty?"

Our painter-friend wheeled around to face me, and I was surprised to see black anger in his eyes. He shouted at me and apparently at everyone else in the gallery, "Beauty is an old-fashioned, idiotic concept. Representational art is dead, killed by the camera, by technologists, and by scientists. We contemporary artists are painting ideas." Then, he pointed at me and yelled, "What do you theoretical physicists know about beauty! Nothing!"

I shrugged my shoulders, and if I hadn't placed friendship above the truth, I would have said, "More than you artists, apparently." I surmised that theoretical physicists talk more about beauty than present-day visual artists. I recalled that even as an undergraduate hardly a class in physics or mathematics went by without the word "beautiful" spoken.

During the summer before my senior year at the University of Michigan, I took Introduction to Theoretical Physics, taught by Professor Otto Laporte. For some odd reason, the class met at 7:00 PM. One night, Laporte walked in with two martinis under his belt and announced, "Tonight gentlemen, I'm going to show you something beautiful." He proceeded to elegantly demonstrate several fundamental theorems about vector spaces. When finished, he stepped back from the blackboard and said, "Isn't that beautiful." One student in the class, the very worst one, asked, "What is so beautiful about that?" Laporte was taken back, and after a moment of silence, asked, "Do the rest of you see that this is beautiful?" We all nodded, and several students replied, "Of course." The professor then turned to the student who was blind to the beauty of vector spaces and told him, "You be quiet. The rest of us see it." Laporte told us in his blunt way that intellectual insight and the apprehension of beauty are not democratic, that poor cultural formation, political and economic ideology, and habituation to lies and ugliness can cut a person off from truth and beauty.

Strangely neither Laporte nor any student in his class could give the characteristic marks of beauty. Except for the worst student, we all recognized mathematical beauty without having ever been taught to. In the terminology of ancient philosophy, our souls by nature were attuned to beauty.

When young, I relished beauty wherever I found it — relativity and quantum physics, the chamber music of Mozart, the great drawings of Picasso. I wanted to be close to heaven in this world and never thought once about what beauty is or why it was so important. Years later, after I saw beauty put to evil ends at Los Alamos National Laboratory and everything in my life was called into question, I set myself the task to discover the marks of beauty, not that I wanted to establish a rigorous definition. I was already persuaded that any formula for beauty is a shadow of the experience of a rose, a Mozart sonata, or the Pythagorean Theorem; however, I believed that if I knew the marks of beauty both my experience of art and my understanding of physics would deepen.

The Beautiful

The beauty physicists talk about is not the product of private or idiosyncratic emotion.[429] Einstein gives the three elements of scientific beauty in a single sentence: "A theory is the more impressive the greater the simplicity of its premises is, the more different kinds of things it relates, and the more extended is its area of applicability."[430] Simplicity, then, is the first element of beauty. The "different kinds of things it relates" means how the theory harmonizes disparate things. Thus, we may label the second element harmony. And the extended applicability is a theory's brilliance; that is, how much clarity it has in itself and how much light it sheds on other things. Simplicity, harmony, and brilliance. Each of these calls for a brief explanation.[431]

Simplicity. There exist today other theories of gravity besides Einstein's general relativity, but because they lack simplicity, none are taken seriously. "Most rival theories are convincingly disproved," observes mathematician Roger Penrose. "The few that remain having been, for the most part, contrived directly so as to fit with those tests that have been actually performed. No rival theory comes close to general relativity in elegance or simplicity of assumption."[432] All rival theories of gravity lack what Werner Heisenberg called "frightening simplicity and wholeness."[433] Without wholeness, or unity, a theory is a collection of disconnected ideas and observations that often borders on the frighteningly ugly.

The principle of simplicity implies completeness and economy. "It is because simplicity and vastness are both beautiful that we seek by preference simple facts and vast facts," mathematician-physicist Poincaré explains.[434] A theory beautiful by this standard must account for all the facts with a few simple principles. A demanding standard, indeed! Heisenberg recalls that quantum theory was "immediately found convincing by virtue of its completeness and abstract beauty."[435]

Harmony. "Without the belief in the inner harmony of the world there could be no science," Einstein declares.[436] Heisenberg defines harmony as the "proper conformity of the parts to one another and to the whole."[437] In any science, a good theory will harmonize many previously unrelated facts, for without the harmony of its parts, a theory lacks unity. Harmony also implies symmetry. There is a pleasing symmetry to all the laws of physics. "Every law of physics goes back to some symmetry of nature," observes physicist John Archibald Wheeler.[438] Heisenberg adds, "The symmetry properties always constitute the most essential features of a theory."[439] Newton's third law is a well-known example of symmetry in physics: To every action, there is always opposed an equal reaction. A different mirror symmetry is found on the subatomic level, where to every kind of particle, there corresponds an anti-particle with the same mass but with opposite characteristics. In fact, this symmetry successfully predicted the existence of many subatomic particles.

Brilliance. A theory with brilliance has great clarity in itself and sheds light on many things, suggesting new experiments. Newton, for example, astounded the world by explaining falling bodies, the tides, and the motions of the planets and the comets with three simple laws. Physicist George Thomson states, "In physics, as in mathematics, it is a great beauty if a theory can bring together apparently very different phenomena and show that they are closely connected; or even different aspects of the same thing."[440] General relativity does precisely that in an elegant and surprising way, as astrophysicist S. Chandrasekhar explains: "It consists primarily in relating, in juxtaposition, two fundamental concepts which had, till then, been considered as entirely independent: the concepts of space and time, on the one hand, and the concepts of matter and motion, on the other."[441] Moreover, general relativity has proven extraordinarily brilliant, shedding light on cosmology and the fate of the universe.

Both general and special relativity, like all theories that possess brilliance, have what Francis Bacon calls "strangeness in the proportion," or fitting surprises.[442] From the principle that the velocity of light is the same for all inertial (non-accelerating) observers, the surprises of relativity — time dilation, length contraction, and the equivalence of inertial mass and energy — follow as inevitable consequences. Einstein's 1905 paper on special relativity is a series of fitting surprises, not unlike what is found in the great music of Mozart and Beethoven.

Music conductor Leonard Bernstein's comments on Beethoven's *Eroica* Symphony apply equally well to Einstein's 1905 paper: "The element of unexpected is so often associated with Beethoven. But surprise is not enough; what makes it so great is that no matter how shocking and unexpected the surprise is, it always somehow gives the impression — as soon as it has happened — that it is *the only thing that could have happened at that moment.* Inevitability is the keynote. It is as though Beethoven had an inside track to truth and rightness, so that he could say the most amazing and sudden things with complete authority and cogency."[443] The truths of relativity, or quantum physics for that matter, are stranger than we imagined, but their inevitability convinces us that Nature is that way. Strip away fitting surprises and beauty is reduced to monotonous unity, mechanical symmetry, and uninteresting predictability. Nuanced broken symmetries in art and science command our attention and force us to contemplate or to seek deeper, underlying truths.

Beauty is not confined to theory; an experiment can possess a simplicity of apparatus, a brilliance that results from the transparency of data analysis, and an outcome that reveals something fundamentally new. The most beautiful experiments transform the way scientists think about Nature.[444]

Until Galileo experimented with rolling balls down inclined planes in the early 1600s, scientists believed terrestrial phenomena lacked mathematical intelligibility. Galileo showed experimentally that the acceleration of a ball rolling down an inclined plane obeys simple mathematical rules. This crucial discovery changed how scientists thought about matter and led Newton to discover the three general laws that govern all mechanical motion. Until 1803, when Thomas Young experimented with light beams, physicists believed all natural phenomena obeyed Newton's laws and thus held that a beam of light was made of material particles. Through a simple experiment, Young observed that two light beams interact like waves, not particles. He divided a light beam in two by passing it over a card, about one-thirtieth of an inch in breadth. A series of light and dark bands resulted, not unlike what happens when a water wave on a lake encounters a square dock post. Young's key experiment altered forever how physicists think about light and eventually led James Clerk Maxwell to discover the equations that describe all electromagnetic phenomena.

The ten paragraphs I wrote on the marks of beauty I circulated to former colleagues at Los Alamos National Laboratory and at the University of Michigan. In the past, over beers and nachos, I heard them voice cultural platitudes: "Beauty is in the eye of the beholder;" "Beauty is very personal;" "Beauty is not objective and thus varies from culture to culture." But I discounted the voice of culture, for I knew that such remarks were contrary to their experience of doing physics and mathematics. My ten paragraphs drew mainly positive responses; the most negative comment was an emotional rant about how I was thinking like a philosopher, not a physicist. Except for this one person, no one objected to the four marks of beauty that I proposed — simplicity, harmony, brilliance, and fitting surprise.

I suspected that all my former colleagues in their youth were shaken and transformed by an experience of some profound beauty. I know I was. In the tenth grade, I was changed forever by Euclid's proof that the prime numbers are infinite, an exquisite proof that surprisingly showed in six lines of text an eternal truth. Until that point in my life, I thought truth did not exist; everything about me changed, the seasons, my body, and people. My experience of the human world was that everything was in flux, sometimes bordering on the absurd. Suddenly, mathematics presented me with one unchanging thing, a timeless truth, demonstrated in an exceedingly beautiful way; the 2,500 years between Euclid and me were of no consequence. My encounter with Euclid was not unique. Bertrand Russell described his first encounter with serious mathematics: "At the age of eleven, I began Euclid, with my brother as my tutor. This was one of the great events of my life, as dazzling as first love. I had not imagined that there was anything so delicious in the world."[445]

Years later, after I left Los Alamos, I was stunned by a colossal failure: Modern science cannot explain why Nature overflows with beauty, both to the eye and to the mind. Leptons, quarks, the chemical elements, the DNA molecule, crystals, snowflakes, sand dollars, hummingbirds, hawks, mice, chipmunks, jaguars, lions, horses, zebras . . . nothing in Nature is not beautiful in some way. (See Figure 22.1.)

Figure 22.1. The logarithmic spiral of the nautilus shell. (Chris 73, Wikimedia Commons.)

The only proposed scientific explanation of this fundamental aspect of Nature is that the origin of beauty is sex. Sigmund Freud reduced beauty to an instinct: "Psychoanalysis, unfortunately, has scarcely anything to say about beauty . . . All that seems certain is its derivation from the field of sexual feeling."[446] In the same vein, Charles Darwin declared that the beauty of present-day animals resulted from sexual selection: "the more beautiful males having been continually preferred by the females."[447] Sexual selection, of course, does not explain why we humans find nightingale songs, pheasant tails, and rainbow trout beautiful. Or, why we delight in the sparkle of a diamond, in the shape of the filigreed wing of a housefly, or in the elegance of Euclid's demonstration of the Pythagorean Theorem.

Darwin acknowledged, "How the sense of beauty in its simplest form — that is, the reception of a peculiar form of pleasure from certain colors, forms, and sounds — was first developed in the mind of man and of the lower animals is a very obscure subject. . . . but there must be some fundamental cause in the constitution of the nervous system in each species."[448] Consequently, Darwin held that the "sense of beauty obviously depends on the nature of the mind irrespective of any real quality in the admired object."[449] But the beautiful equations of general relativity, which can be written in the palm of a hand, "contain" such "objects" as black holes, the Big Bang, and the fate of the universe.

Darwin was too great a scientist not to be enthralled by the abundant beauty of Nature. "Delight itself," he said, "is a weak term" to describe the deep pleasure a naturalist experiences in a Brazilian rain forest.[450] In the concluding words of *The Origin of Species*, Darwin invoked beauty as a proof that his theory is true: "from so simple a beginning endless forms most beautiful and most wonderful have been, and are being evolved."[451] But to speak of the "most beautiful and most wonderful" organic forms is nonsense, if beauty "depends on the nature of the mind irrespective of any real quality in the admired object." To get out of this contradiction, beauty must be a real property of Nature.

The True

The answer to the question "Why does Nature abound with beauty?" begins where Darwin ends, that is, with a working principle of science: "You can recognize truth by its beauty." Physicist Fred Hoyle relates that when he was young and not many years into research, his thesis advisor Paul Dirac once told him, "Hoyle, you are much too empirical. Look more closely at the mathematical structure of what you are doing." When Hoyle asked what he should look for, Dirac answered, "You have to learn to recognize what is beautiful."[452]

Beauty, then, is a means of discovering scientific truth. "In every epoch in the history of science," physicist Louis de Broglie observes, "aesthetic feeling has been a guide that has directed scientists in their research."[453] James Watson, for instance, describes in his book *The Double Helix* how beauty guided Francis Crick and him to discover the structure of DNA: "So we had lunch, telling each other that a structure this pretty just had to exist." And "almost everyone . . . accepted the fact that the structure was too pretty not to be there."[454] According to Richard Feynman, in science, "you can recognize truth by its beauty and simplicity."[455]

The more something strays from simplicity, harmony, and brilliance, the less beautiful it becomes *and* the less intelligible. A circle is manifestly intelligible; we recognize it immediately and can define it. But a circle with an ever so slight asymmetric irregularity lacks intelligibility — it does not even have a name. Without simple, symmetric patterns, Nature would not be intelligible; without the brilliance of underlying, universal causes, Nature would not be intelligible. If Nature were ugly, science would not exist. But Nature abounds with beauty, and that is why it is supremely intelligible.

The intelligibility of Nature is astounding. As we noted in Chapter 1, except for the vague, mystical theories of Pythagoras, the ancients never envisioned that terrestrial phenomena obey mathematical laws. Earthly motion is bafflingly complex, and unlike the Sun and Moon, the prediction of the future of anything on Earth seems impossible. It is far from obvious that a projectile on Earth traverses a parabolic path, that the colors of the rainbow exhibit a simple mathematical pattern, and that underlying the bewildering, everyday world of chemical changes, such as the rusting of iron or the burning of sugar, are simple, mathematical rules. Every herder and farmer, since the domestication of animals and the inception of agriculture, knew an offspring resembles its parents, but Gregor Mendel was the first person to discover that the laws of inheritance are written in terms of simple, whole numbers.

Nature is not just sort of mathematical, it is staggeringly so. Newton verified the law of gravity with a four percent accuracy; subsequently, the law was determined to be accurate to a ten-thousandth of a percent. Modern physics experiments are unbelievably accurate. Consider the magnetic moment of an electron, which is something like the force of a tiny magnet. Its value, in certain units that are not important for the present discussion, is around 1. The experimental value is 1.001159652188, with an uncertainty of about 4 in the last digit; the value calculated from the theory of quantum electrodynamics is 1.00115965214, with an uncertainty of about 3 in the last digit. This agreement between theory and experiment is like the measured and the calculated distance between New York and Los Angeles agreeing within the width of a human hair.[456] "The numerical agreement between theory and experiment here is perhaps the most impressive in all science," Steven Weinberg declares.[457] The mathematical nature of the universe is truly amazing.

The uncanny accuracy with which theoretical physics describes Nature led Einstein to say, "The most incomprehensible thing about the world is that it is comprehensible."[458]

The Good

If existence were not good, we would be indifferent about death; we would greet a diagnosis of a fatal disease with a yawn. But we have developed medical technologies to fight disease and to extend longevity. Our bodies, like all organisms on Earth, maintain life as long as possible; physically, emotionally, and intellectually, we desire life to last forever. On our best days, we awake and tell ourselves a profound truth: "It is great to be alive."

We know that this good existence is not forever. Even considering the end of the universe saddens some physicists. Theorist Edward Witten, contemplating how with the Big Freeze the universe ends like a wisp of smoke, said such a fate "is not very appealing." Astrophysicist Glenn Starkman concurs that such a universe "would be the worst possible universe, both for the quality and quantity of life," and his colleague Lawrence Krauss adds, "All our knowledge, civilization, and culture are destined to be forgotten. There's no long-term future."[459]

Divine Mind

Physicist Eugene Wigner succinctly describes the three miracles that science relies upon: "The miracle that the human mind can string a thousand arguments together without getting itself into contradiction . . . [and] the two miracles of the existence of laws of nature and of the human mind's capacity to divine them."[460]

Wigner calls these "miracles" because no scientist has even a bad idea to offer about how mind could evolve out of matter and then come to comprehend the deep mathematical structure of matter. Not one neuroscientist has proposed even a goofy mechanism for how a complex arrangement of neurons in the human brain can give rise to the Ricci tensor of general relativity or to the Hilbert spaces of quantum mechanics, nor has any evolutionary psychologist shown how the Schrödinger equation resulted from the reproductive success of our ancestors on the African savannah.

Once the truth that no arrangement of matter can produce mind is acknowledged, then the most straightforward explanation of these three miracles is that the universe results from the Divine Mind and that in some way the human mind is akin to the Divine Mind. Reflecting on the mathematical structure of the universe, Dirac suggests, "It seems to be one of the fundamental features of nature that fundamental physical laws are described in terms of a mathematical theory of great beauty and power, needing quite a high standard of mathematics for one to understand it . . . One could perhaps describe the situation by saying that God is a mathematician of a very high order, and He used very advanced mathematics in constructing the universe."[461]

The Greek word for beautiful is *kalos*, which bears an uncanny resemblance to the verb *kaleô*, "to call." A beautiful object calls us to it, and beyond to its maker. In this way, the beauty of a red rose, of a soaring hawk, or of a natural law calls us to the Divine Mind. Pointing to the deepest aspects of the human soul, Socrates said a lover of wisdom questing for universal beauty will find himself or herself "ever mounting the heavenly ladder, stepping from rung to rung — that is from one to two, and from two to *every* lovely body, from bodily beauty to the beauty of institutions, from institutions to learning, and from learning in general to the special lore that pertains to nothing but the beautiful itself — until at last the lover of wisdom comes to know what beauty is."[462] The quest for beauty ends when the lover's eyes are opened to gaze upon Divine Beauty in true contemplation.

The universe has precisely the properties that we would expect if it were brought into existence by the Divine Mind — the universe is supremely intelligible, exceedingly good, and abundantly beautiful. If we take the universe to be an expression of the transcendent, then the Divine Mind, or God, is intelligible, good, and beautiful.[463]

23 God in Modernity

From my habit of proceeding by way of contrast, I first turned to the high point of Christianity in both the East and West before attempting to understand the concept of God in Modernity. According to patristic tradition, God can be known in two ways. Cataphatic, or rational, knowledge attained through *dianoia*, defines God by positive statements; apophatic knowledge is direct experience of God by *nous*, although such knowledge cannot be expressed in words. God, however, is *not* known in his essence through either of these. *Dianoia* understands God as the creating and sustaining Cause of the world as argued in the preceding chapter. *Nous* gives a direct experience of His mystical presence, which surpasses the rational knowledge of God as Cause and transcends all possibility of definition. These two kinds of knowledge are neither contradictory nor mutually exclusive; rather, they complement each other. Apophatic knowledge of God, clearly, needs elaboration.

God is not identical with any of his named qualities given by rational knowledge; neither eternal nor infinite nor simple, God is beyond all categories of human thought. The essence of God cannot be named; thus, God is the Unnamable. In his book *The Divine Names*, Pseudo-Dionysius the Areopagite[464] tabulates 52 names of God used in the Old and New Testaments.[465] A few are God of gods, Holy of holies, Cause of the ages, the still breeze, cloud, and rock. Pseudo-Dionysius insists that God is *not* Mind, Greatness, Power, or Truth in any way we can understand. God "cannot be understood, words cannot contain him, and no name can hold him. He is not one of the things that are and he is no thing among things."[466] Pseudo-Dionysius says the path to God is through unknowing: "The more our words are confined to ideas we are capable of forming . . . we shall find ourselves not simply running out of words but actually speechless and unknowing."[467] Such speechlessness opens the possibility that *nous* may directly experience the mystical presence of God.

Subsequent Church Fathers whole-heartedly embraced the insights of Pseudo-Dionysius. John of Damascus (c. 675–749) wrote, "God does not belong to the class of existing things: not that He has no existence, but that He is above all existing things, nay even above existence itself."[468] St. Peter of Damascus (twelfth century) concurred: God "is always beyond goodness, righteous, all-wise, all-powerful, unconquerable, dispassionate, uncircumscribed, infinite, unsearchable, incomprehensible, unending, eternal, uncreated, invariable, unchanging, true, incomposite, invisible, untouchable, ungraspable, perfect, beyond being, inexpressible, inexplicable, full of mercy, full of compassion and sympathy, all-ruling, all-seeing."[469]

St. Maximos the Confessor (580–662) pointed to the tension between cataphatic and apophatic knowledge of God: 1) "We do not know God from His essence. We know Him rather from the grandeur of His creation and from His providential care for all creatures. For through these, as though they were mirrors, we may obtain insight into His infinite goodness, wisdom, and power;"[470] and 2) "If you are about to enter the realm of theology, do not seek to descry God's inmost nature, for neither human intellect nor that of any other being under God can experience that."[471]

A crucial problem in the patristic tradition was how to combine two statements about God that appear contradictory: 1) We can know God, and 2) God is by nature unknowable.

St. Gregory Palamas is credited with the resolution of this paradox; we know the *energies* of God, not His *essence*: "Not a single created being has or can have any communion with or proximity to the sublime nature [of God]. Thus, if anyone has drawn near to God, he evidently approached Him by means of His energy."[472] Here, Gregory distinguishes between God's essence, or substance (*ousia*), and His activity (*energeia*) in the world. The energies of God are experienced as Divine Light, such as the light of Mount Tabor or the light that blinded St. Paul on the Road to Damascus. The Church Fathers, however, warned that for most devout Christians, visions and heavenly voices do not come from God but from a fevered imagination and are distractions in the spiritual quest.

Modern spiritual directors issue the same warning about imagined experiences. Thomas Merton, for example, told his novices that "since God cannot be seen or imagined, the visions of God we read of the saints having are not so much visions *of* Him as visions *about* Him; for to *see* any limited form is not to see Him."[473]

Thomas Aquinas (1225–1274), the most rational of theologians, also reached the conclusion that God is beyond the comprehension of *dianoia*: "It is because human intelligence is not equal to the divine essence that this same divine essence surpasses our intelligence and is unknown to us: wherefore man reaches the highest point of his knowledge about God when he knows that he knows Him not, inasmuch as he knows that that which is God transcends whatsoever he conceives of Him."[474] The terms good, wise, and just do not signify the essence of God "perfectly as it exists in itself, but as it is conceived by us."[475] Following Pseudo-Dionysius the Areopogite, Aquinas says wisdom can be applied to God in three ways: 1) God is wise, since in Him there is a likeness to wisdom; 2) God is *not* wise, since the wisdom in Him we cannot understand and name; and 3) We ought to say God is supra-wise, since He transcends the wisdom we indicate and name.[476]

Aquinas points to the three-step dialectical method of spiritual enlightenment introduced into Christianity by Pseudo-Dionysius. The novice spiritual seeker first affirms what God is; he says God exists, God is Wisdom, and God is One. As his spiritual life deepens, the novice realizes the absurdity of conceiving God as a superhuman, and he denies that God exists, that God is Wisdom, and that God is One. But the way of denial is just as faulty as the way of affirmation. The spiritual seeker then discovers that God is "beyond every assertion" and "beyond every denial."[477] The third dialectical step thus is a kind of aphasia, the loss of language brought about by intense intellectual analysis of the nature of God, who falls "neither within the predicate of existence nor non-existence."[478] The spiritual seeker now "speechless and unknowing ... [is] supremely united to the completely unknown by an inactivity of all knowledge, and knows beyond the mind by knowing nothing."[479]

The Eastern Fathers vigorously fought the temptation to replace the experience of the unfathomable depths of God with philosophical concepts. Unlike the patristic understanding of God where the substance of God is unknowable and such named attributes of God as good, wise, and just are defective and imperfect, René Descartes (1596 – 1650) introduced the modern concept of God, the highest being among beings, a larger and more powerful version of humans, knowable by rational arguments, and literally possessing the highest virtues of humankind.

In his *Meditations on First Philosophy*, Descartes uses logic to demonstrate the existence of God. He first observes that he has ideas of the eternal, infinite, immutable, omniscient, and omnipotent; he then asks whence came these ideas, since he is none of these things. Because he cannot be the author of these ideas, they must come from a being outside of him who is eternal, infinite, immutable, omniscient, and omnipotent, that is, from God. Descartes concludes that he is not surprised that God in creating him placed the idea of God within him, like "a mark of the craftsman imprinted on his work."[480]

This modern God resembles the gods of primitive religions; He is a superhuman open to any criticism of heroic warriors, secular rulers, and ordinary mortals. The Homeric gods, for instance, were essentially immortal men and women. The gods, although vastly more powerful than the mortals, possessed every human emotion and appetite. Many committed such human vices as treachery, deceit, and sexual excess. Just as Socrates subjected the Homeric gods to philosophical critique readily applied to human rulers, some modern atheists condemn God as a despot enslaving human beings with a moral code impossible for any mortal to live up to. Christopher Hitchens expresses this sentiment in his introduction to *The Portable Atheist: Essential Readings for the Nonbeliever*: God is the "utter negation of human freedom. . . . Who wishes that there was a permanent, unalterable celestial despotism that subjected us to continual surveillance and could convict us of thought-crime, and who regarded us as its private property even after we died?"[481]

Descartes envisaged the cosmos as a clockwork mechanism and God as its designer and maker, a philosophical image embraced by deists of the Enlightenment. The supposed Divine Craftsman fashioned mindless matter into a cosmic machine run by elegant mathematical laws, although now He is outside the physical universe and does not interfere with its operation. In antiquity, such a god was the Demiurge, a god amongst gods, not the source of all existence.[482] In early Modernity, such a god was the center of Deism, a rational religion held by such Founding Fathers of America as Thomas Paine (1737–1809), Benjamin Franklin (1743–1826), and Thomas Jefferson (1711–76).

Most Christians in the twenty-first century do not see God as the ultimate mystery and believe the words they use to express God's attributes are literally true. Theologians "capture" God with their true doctrines; biblical students "contain" God within the Holy Book; churchgoers "understand" God, every Sunday, inside the one, true Church. Such Christians believe they live near God and know Him intimately, much like they live near and know their neighbors. In Sunday school, boys and girls learn that God is like the nice man next door, but He also happens to be the truant officer of the universe. No person, however, knows the hidden, interior life of another person, much less the essence of God. A danger inherent in modern Christianity is that the "understanding" of God becomes an empty idol, a false god, a mere human projection, a danger recognized by the patristic father Gregory of Nyssa (c. 335–c. 395): "Every thought and every denning conception which aims to encompass and grasp the divine nature is only forming an idol of God, without declaring Him as He truly is."[483]

Today, organized religion is irrelevant to how most people live, especially to the young in the United States and Western Europe. The prevailing political and economic order of democracy and capitalism has no place for religion. Democratic equality erodes the hierarchical authority of every institution, including the Church; capitalism focuses life on material prosperity, away from eternal salvation; science, the sole intellectual authority, denies the history of humankind given in the Bible. With the cultural emphasis on the individual, the Christian virtues of humility, obedience, and self-sacrifice vanish in favor of self-achievement, getting ahead in the rat race, and making a name for oneself. Few Christians today see themselves as pilgrims journeying through this life to the next, avoiding the attachment to worldly things and hoping to evade the snares set by the Devil.

Not surprisingly, the World Values Survey (2000) discovered that in 12 major European countries, 38 percent of people say they never or practically never attend church. In some areas of France, Sweden, and the Netherlands, church attendance is less than 10 percent. The Swedish government reported that 85 percent of Swedes are church members, but only eleven percent of women and seven percent of men go to church regularly. By contrast, in the

United States, only 16 percent say they rarely go to church.

Long before the obvious decline of religion in the twenty-first century, the Nation-State had absorbed the allegiance of the masses and took over many of the former functions of the Church, such as education and establishing laws that govern marriage and divorce. In wartime, opposed Nation-States do call upon religious leaders to proclaim God is on their side and to pray for victory. In peacetime, however, priests, ministers, and rabbis are forbidden by democratic consensus to advocate moral restraints on free markets or to interfere with lawmaking, apart from such moral issues as abortion and same-sex marriage. In present-day America, churches and synagogues, except for providing soup kitchens and homeless shelters, have virtually no role in political or economic life. Any institution with little or no influence on everyday living soon disappears, as the medieval guilds and the extended family in America did.

Judeo-Christian Atheism

The modern world is founded on atheism, not the atheism of antiquity, such as that expounded by Lucretius (c. 99 BC–c. 55 BC), but — and at first this may sound strange — a Judeo-Christian atheism, which is not a contradiction in terms. Judeo-Christian atheism denies the existence of God, agrees with Lucretius that in matter "lies the sum of all created things,"[484] and yet borrows key elements from the Hebrew Bible and the New Testament.

Copernicus learned from Genesis that "the Best and Most Orderly Workman of all"[485] created the cosmos; consequently, the cosmos must be intelligible and beautiful. The second element that Copernicus drew out of Genesis is that since man and woman were made in the image of God, natural philosophers could discover "the truth in all things, in so far as God has granted that to human reason."[486] Four hundred and seventy-five years after the publication of Copernicus' *On the Revolutions of the Heavenly Spheres*, physicists and philosophers, now inhabiting a Godless universe, assume that nature is intelligible, but offer no explanation for this central tenet of science, nor do they explain why *Homo sapiens* can apprehend the hidden, highly abstract mathematical laws of nature.

Christianity demystified nature banished the spirits, demons, gods, and goddesses of antiquity, and denied the sacredness of plants and animals. With nature no longer sacred, Francis Bacon (1561–1626) could establish the goal of modern science, a restoration to the Garden of Eden, so that once again man would command nature. He envisaged a "great mass of inventions"[487] that would contribute to the "relief of man's estate,"[488] a way to treat some diseases, bring about modest improvements of health, and perhaps increase physical comfort and pleasure — "in some degree [to] subdue and [to] overcome the necessities and miseries of humanity."[489]

Nationalism, in contrast to patriotism, draws three elements from the Hebrew Bible — the chosen people, the Covenant, and the Messianic expectancy.[490] Moses told the Children of Israel, "The LORD your God has chosen you to be a people for his own possession, out of all the peoples that are on the face of the earth" (Deuteronomy 7:6). The chosen people were united to God, forever: "And I will establish my covenant between me and you and your descendants after you throughout their generation for an everlasting covenant, to be God to you and to your descendants after you" (Genesis 17:7). God would lead the chosen people to a Messianic Age, a future time without poverty, civil strife, and war. "They shall beat their swords into plowshares and their spears into pruning hooks; nation will not lift sword against nation, neither shall they learn war any more" (Isaiah 2:4).

The Puritans believed they were the new chosen people of God, and that belief has remained throughout the history of the United States. Thomas Jefferson declared in his Second Inaugural Address that Americans were a chosen people to whom God had shown His favor, for He "led our fathers, as Israel of old, from their native land and planted them in a country flowing with all the necessaries and comforts of life; who has covered our infancy with His providence and our riper years with His wisdom and power."[491]

At first glance, the modern world drew nothing from the New Testament, from the Gospel of Love. Neither the Nation-State nor the free market resides in love or friendship. As we saw, John Locke (1632–1704) supplied the philosophical foundation for both modern democracy and free-market capitalism in his book *Second Treatise of Government*, and James Madison (1751–1836) concurred in *The Federalist Papers*. Both the philosopher and the future president of the United States held that in a state of nature, each individual, alone, without love or friendship, "full of fears and continual dangers,"[492] encounters hate and enmity. What forces individuals to form civil society is love of self and self-preservation. The free market, too, is based on self-love and the disregard of the good of others. The theoretical underpinnings of the Nation-State and capitalism reject the Gospel of Love, unlike the ancient Greek understanding that the polis was held together by *philia* (friendship).[493] The modern world totally denies Christianity in the sense that love is banished from public life and is confined to the private sphere of friends and family.

In the New Testament, love and freedom are joined forever. St. Paul tells the Galatians that "the whole law is fulfilled in one word, 'You shall love neighbor as yourself'" means that "you were called to freedom."[494] For a person to love God and neighbor, he or she must be free; otherwise, a person is a slave of custom, law, or another. Each person — master or slave, rich or poor, old or young, man or woman — was given the freedom to choose to be obedient to the commandment of love. God does not desire slaves but lovers.

In present-day America, personal freedom — not political liberty — is embraced as *the* central aspect of human life; freedom and obedience are thought to be incompatible. Besides freedom, Modernity drew from the New Testament a buoyant optimism, the positive emotions accompanying the belief in the dawning of a new age heralding a bright future, where all the current ills of human life — poverty, illness, and strife — would be overcome. The course of history, however, was understood not as the continuing dialogue between God and man but as the progressive emancipation of the individual from the tyranny of hierarchical social structures that in the past oppressed the vast majority of men and women. In the future, when freed from all oppressive regimes, each individual will be free to choose her own end so that she will be obedient only to herself.

Judeo-Christian atheism, then, holds that only matter exists, that the cosmos is governed by intelligible laws, graspable by men and women, and that through the command of nature, humankind will achieve its final historical destiny, Paradise on Earth. Such a view of human nature and history is a jumble of contradictions: Men and women are free agents, yet slaves of matter; technological innovations produce happiness, yet Hiroshima, the Gulag, and the Holocaust would have been impossible without science and engineering; all gods are illusions, yet humans worship the Nation-State; men and women are driven by self-interest, yet citizens are called upon to sacrifice their lives and well-being for the good of the New God; the possession of material goods make a person happy, yet with abundant riches, Americans are lonely and discontent; each person is King of the Castle, the sole judge of what is true, good, and beautiful, yet an insignificant speck in the cosmos.

24 Death and Blind Hopes

Prometheus was the one Olympian god to rebel against Zeus' plan to wipe out the human race and create another. In this Greek myth, as told by Aeschylus, humans lived like ants in sunless caves, possessing neither science nor technology. With the ability to foresee the future, the fear of death paralyzed them. Prometheus took pity upon these wretched creatures and "caused them to cease foreseeing death," lodged "blind hopes" in their hearts, stole fire from Mount Olympus and gave it to humans so they would learn the arts for a good life in this world and thus further blind themselves to the inevitable death that awaits all mortals.[495]

We die; everything perishes.

Yet, we believe the United States of America will last forever; we hope to preserve our memory by founding a college, a business, or a family; we dream of immortalizing ourselves by setting a sports record, making a significant scientific discovery, or writing the great American novel. But the Anasazis, the Hittites, and the Sumareans remain only in name; barely legible inscriptions on granite headstones proclaim the deceased is "gone, but not forgotten;" and, Jack "Gentleman" Johnson, Friedrich August Kekule, and Theodore Dreiser, giants in their own days, are now historical footnotes. Nothing escapes time; everything flows to oblivion.

The eleventh-century Chinese scholar-poet Su Tung-po compared all human endeavors to the footprints left by geese on snow:

> To what can our life on earth be likened?
> To a flock of geese,
> Alighting on the snow.
> Sometimes leaving a trace of their passage.[496]

The writer of *Ecclesiastes* preaches that everything in this world — wisdom and folly, riches and poverty, happiness and grief — "all is vanity and a striving after wind." A man dies and nothing remains.

> Vanity of vanities, says the Preacher,
> vanity of vanities! All is Vanity.
> What does man gain by all the toil
> at which he toils under the sun?
> A generation goes, and a generation comes,
> but the earth remains for ever.[497]

Most of us do not despair over our dreadful situation. In the ancient Indian epic poem the *Mahabharata*, the sage Yudhisthira is asked, "Of all things in life, what is the most amazing?" Yudhisthira answers, "That a man, seeing others die all around him, never thinks that he will die."[498] In the East and the West, men and women use amusement and fantasy to divert their attention from the reality of death. Philosopher-mathematician Blaise Pascal observed in his day, the early seventeenth century, that the chief diversions were the hunt and the dance. A man whose mind is occupied with where to rightly put his feet cannot reflect upon the inevitable death that awaits him.[499] In our democratic era, the diversions readily at hand to us far surpass those available to seventeenth-century French aristocrats. With the touch of a finger, we change channels on the TV, select a different track on a CD, or go to a new Website on the internet.

What we hate most of all is silence and solitude; home alone, the TV set drones away in the background, even when we are not watching it. We get into the car by ourselves, and our hand automatically reaches to turn on the radio. We avoid every opportunity to reflect upon where we come from, what we are doing here, and where we are going. When reality intrudes, we reach for the psychotropics and thank Pfizer, the maker of Zoloft, for taking the edge off life.

In the modern world, particularly in America, we blind ourselves to illness, aging, and dying. The critically ill are moved to intensive care units, where they are hidden beneath cables and tubes connected to monitors and respirators; the old and infirm with cognitive problems are housed in memory-care facilities and are often confined to a single room with a flat screen TV and artificial flowers; the dying spend their last days in a hospice center, staffed by professionals, all strangers.

The Buddha unflinchingly stared Death down. He preached in his first sermon to five ascetics, his old companions, in the Deer Park at Isiptana near Benares that human existence is inseparable from suffering (*dukka*), that the cessation of suffering occurs by extinguishing craving, and that the liberation from craving results from ardently following the Eightfold Path.[500]

Dukka follows from the most fundamental principle that Buddha taught: the impermanency of all compound things, an indisputable principle. Civilizations rise and fall; species of plants and animals come and go; continents drift and produce mountain ranges; wind and water erode rock and level mountains. Twentieth-century cosmologists discovered that the universe, itself, is destined to end with the Big Freeze, a cold eventless state of electrons, neutrinos, antielectrons, and antineutrinos. We are born, walk around for a while, and then disappear. Everything and everyone we love changes inevitably, decays, dies, and vanishes. Nothing lasts. All this is indisputable, although unlike the Buddha we blind ourselves to this reality.

Surprisingly, many American Buddhists are experts at denying death, unlike their Master. Buddhism in our culture verges on being a bourgeois amenity, like Whole Foods, where the production of fruits, vegetables, and meat is sanitized, hidden from view. Robert Wright, in his best-selling book *Why Buddhism is True: The Science and Philosophy of Meditation and Enlightenment*, translates *dukka* as "unsatisfactoriness." Life is a series of disappointments; "getting the next job promotion, or getting an A on that next exam or, eating that next powdered-sugar doughnut" does not bring "eternal bliss,"[501] nor suffering like watching your spouse die of brain cancer. Meditation, "an essentially therapeutic endeavor — a way to relieve stress or anxiety, cool anger, or dial down self-loathing just a notch — can turn into a deeply philosophical and spiritual endeavor,"[502] although Wright confesses that *nirvana* remains a mystery to him and that so far meditation has made him only somewhat less irritable. In effect, Wright transforms Buddhism into a well-being commodity, drained of all reference to death or the transcendent.

Many technologists in Silicon Valley are committed to the blind hope that death is a technical problem that scientists will soon solve. Google contributed to founding Singularity University, which is not a university and where Singularity refers to the tipping point "in the history of the race beyond which human affairs, as we know them, could not continue," according to John von Neumann, the greatest mathematician of the twentieth century.[503]

Von Neumann gave no details of what the Singularity in human life might be, but numerous science-fiction writers, futurists, computer scientists, physicists, and mathematicians have. Vernor Vinge, a mathematician and science-fiction author, predicted, in 1993, that "within thirty years, we will have the technological means to create superhuman intelligence. Shortly after, the human era will be ended."[504] Inventor and futurist Ray Kurzweil placed the Singularity at 2045 when the explosive increase in computing power will have produced superintelligence that transcends the "limitations of our biological bodies and brains."[505] Computer scientist Jaron Lanier claims that in the new religion of information, humans will "become immortal by getting uploaded into a computer,"[506] what believers in the Singularity call "digital ascension." The blind hope is that humans and computers will soon merge, making old age and death disappear.

Historian Yuval Noah Harari in his book *Homo Deus* discusses Google's anti-aging program and claims that the company "probably won't solve death in time to make Google co-founders Larry Page and Sergey Brin immortal."[507] When Brin, 44 at the time, heard this, he said, "Yes, I was singled out for death; no, I'm not actually planning to die."[508]

Two good friends of mine, both believers in the Singularity and adherents to a calorie-restricted diet, scoffed at my assertion that the pursuit of science-based immortality is akin to the search for a perpetual motion machine — no one has ever beaten entropy — things inevitably fall apart.

Most of us mortals are not believers in the Singularity and are more like Susan Sontag, a great writer, a public intellectual, and an extremely erudite thinker. She was diagnosed with metastatic breast cancer, in 1975, at the age of 42, and elected to have, according to her son, David Rieff, the "most radical, mutilating treatment"[509] for her stage IV cancer. Her elected course of treatment saved her life, and she "became a militant propagandist for more rather than less treatment,"[510] despite that more medical intervention for cancer can be exceedingly harmful, causing new cancers, heart attack, and stroke.

Almost thirty years later, Sontag was diagnosed with myelodysplastic syndrome (ADS), a cancer in which immature blood cells in the bone marrow do not mature. Even today, ADS has no satisfactory treatment and usually converts to acute myelogenous leukemia that eventually kills the patient.

Sontag loved life — "there was nothing she did not want to see or do or try" — and was "unreconciled to mortality."[511] She sought a reprieve from her death sentence and may have believed that "death is somehow a mistake, and that someday that mistake will be rectified," according to Rieff.[512] "What she wanted was survival, not extinction — survival on any terms."[513]

Sontag convinced herself that the best chance for her survival was a bone marrow transplant at the Fred Hutchinson Cancer Research Center in Seattle. The transplant failed, her leukemia returned, and she died "slowly and painfully."[514]

Rieff knew his mother had a "horror of cremation,"[515] and he had her buried in the Montparnasse Cemetery in Paris. On reflection, he writes, a "black polished slab covers the bones and whatever else now remains of the embalmed corpse that was once an American writer named Susan Sontag, 1933–2004."[516]

Like Sontag, many of us in the modern world fear "extinction above all else,"[517] so how strange to us is Socrates' claim that "true philosophers make dying their profession, and that to them of all men death is the least alarming."[518] Socrates argued that if a person fixes his attention on the material world, he falls "prey to complete perplexity and uncertainty"[519] for the senses can report that the same object is both hot and cold, big and small. Consequently, he advocated a philosophical training that "consists in separating the soul as much as possible from the body, and accustoming it to withdraw from all contact with the body and concentrate itself by itself."[520] After such a philosophical training that aims at the direct experience of the eternal, the "soul can have no grounds for fearing that on its separation from the body it will be blown away and scattered to the winds, and so disappear into thin air, and cease to exist."[521]

On the day of his forced suicide, many of Socrates' young friends arrived at his cell early in the morning to spend the last day with the Master. Socrates saw that all his followers were afraid of death, and he spent the final hours of his life consoling them! He attempted to cast a "magic spell"[522] over his disciples to charm away their fear of death. If he could show them that the soul is immortal and what awaits it after death is a life more abundant, then in a sense he would have slain death for them. (See Figure 24.1.)

Figure 24.1. *The Death of Socrates* by Jacques-Louis David. Metropolitan Museum of Art, public domain. Wikimedia Commons.

He explained to them that through philosophical training, their souls would experience the invisible, the divine, and the timeless, and that such experience inspires confidence in the immortality of the soul. His young friends, however, had doubts and demanded a logical proof that the soul is immortal. Socrates must have known that such a proof would never allay the fears of his friends; nevertheless, out of love, he proceeded to spin a web of logic to dispel the fear of death from his young friends, all the time directing them to the deepest experiences of the interior life.

Socrates used philosophy to wind his way into the labyrinth of human existence to slay the monster — death — and thus to rescue Athenian youth. At his trial, he had refused to save himself from death, and while in prison, he did not permit his wealthy friends to arrange for his escape. In effect, Socrates offered himself as a sacrifice to death in order to teach publicly that the soul is immortal.

Many of us in the modern world are awed that Socrates spent his last day on Earth as he spent the days before, pursuing truth with his friends and looking out for their welfare. He calmly drained the cup of hemlock in one breath, prayed to the gods that his "removal from this world to the next may be prosperous," and asked his friend Crito "to offer a cock to Asclepius," the god of healing; thus died the "bravest and also the wisest and most upright man" in Athens.[523]

In Modernity, self displaces soul, because the basic unit of democracy, capitalism, and the Nation-State is the isolated, autonomous self, as we saw in Chapter 5. Consequently, Socrates' discussion of the nature of the soul is mainly of historical interest to most of us; we what to know if the self is immortal.

While ancient theologians such as Augustine and Aquinas spoke of the immortality of the soul, New World Christians proclaim that the self is immortal. When a bereaved self asks a priest or pastor, "Will I see my loved one again?", the answer is invariably "Yes," with the implication that the desires, habits, and memories of the loved one are either immortal and live on now or will be resurrected in Christ. C. S. Lewis, a New World Christian apologist, even argued (hoped or believed) that his favorite dog would be resurrected with him.[524]

Margaret Guenther, retired director of the Center for Christian Spirituality of the General Theological Seminary of the Episcopal Church, imagines, "Maybe the next life will be a feast for the mind, like great expanses of time in the main reading room in the Library of Congress, only with good lighting and comfortable chairs. Maybe it will be bountiful, like the homecoming picnics at the Martin City Methodist Church, where my father worshiped as a boy." She confesses, "Sometimes I play with the idea that I will see my grandfather, whom I loved deeply and who died when I was nine, and meet my German grandparents for the first time. . . . Maybe I can have a beer with Meister Eckhart or crochet and chat with Dame Julian of Norwich, while she sews on humble garments, suitable for anchorites."[525]

Implicit in Guenther's picture of the next life is her answer to the most fundamental question a person can ask — "Who am I?" Guenther gives the common answer — I am my memories, a view that does not hold up to scientific or philosophical examination. Recall that neurologist Oliver Sacks reported that a man under his care had suffered a sudden thrombosis in the posterior circulation of the brain, which caused the immediate death of the visual parts of the brain. The patient had lost all visual images and memories yet had no sense of loss. An entire lifetime of visual experience had been erased from memory in an instant.[526]

The visual memories stored in a person's brain are nonmaterial, but they do not exist separate from his brain; the same is true for all other perceptual memories. My memory of winning the eighth-grade math prize in 1950 at the Middle School in Union Lake, Michigan will die with my brain, as will my memory of crossing frozen Wilkins Pond during a full moon in mid-winter of the same year. What is true of memory is also true of imagination. My image of myself — a gypsy outsider — will perish with my death. All my acquired emotional habits, such as the fear of dogs and the love of Bach and Mozart, depend upon brain physiology. Even discursive reason, which moves in time, seems perishable.

Through such reflection, my intellect uncovered a terrible truth — my mortality. I then concluded that the mortality of George Stanciu was a calamity that rendered human life meaningless. To ignore or to forget the reality of death, to force this unbearable truth from my mind, I often turned to a sensual life, to amusing diversions, or to other forms of self-narcosis. What kept me from firing a bullet into my brain was that deep down, I thought that I had possibly made an error in my analysis of who I am.

The Self: A Cultural Construct

I was not born speaking English or Romanian, nor did I know Newton's three laws or the Preamble of the U.S. Constitution. I was not born with a self; even when two years old, there was no Georgie Stanciu. Child psychologists have observed that around 19 months, a child begins to use the words "my," "mine," and "me" and his name with a verb — "Georgie eats."[527] By 27 months, self-reference is common, although the child is not telling the parents *who he is*; that requires a narrative. Between three to five years of age, autobiographical memory emerges, and the development of a unique personal history begins.[528] Even at this young age, the self-narrative pattern that emerges depends upon culture.

I, of course, was born into a world of complex social relations; others instructed me how to behave and taught me what was important in life; in effect, the world gave me a self, assigned me a unique node in a social web. My parents and relatives taught me that I was part of a Romanian community with indissoluble obligations to others. I can still hear my father telling me that at times the other guy needs your help.

At school, the lessons I received contradicted my father's teaching. In the third grade, I sat at my desk, in a row of identical desks; mine was the farthest from the teacher, who sat at a large desk at the front of the classroom. Each one of us occupied a small cubicle with invisible walls. When my best friend, Joey Prinko, reached across the aisle separating our respective rows of desks to hand me a pencil or a crayon, the teacher yelled at him and told him to keep his hands home.

Like every person on the planet, I fashioned my desires, achievements, losses, and experiences with others into a coherent whole through a self-narrative. I called my self into existence, as others did, through language. "George

Stanciu" merged the Romanian and American aspects of my childhood into the story "Gypsy Outsider," a narrative that included bits and pieces taken from literature, pop music, and the big screen. I shamelessly stole the plot of *Shane*, my favorite boyhood movie. Shane comes from nowhere, has no family or last name, possesses his own moral code, needs no help from anyone, and rescues the cowardly townspeople from the "bad guys," and then rides off into the sunset. He is the picture of independence, the embodiment of the isolated, autonomous individual.

From early childhood on, I repeatedly told my self-narrative, adding layer upon layer of storytelling, incorporating new experiences into my story, and often embellishing past events to such a degree that they become distorted or even fictitious. I developed intense attachments to certain personal stories, revisiting them again and again, for weeks, months, and even years. In this way, "George Stanciu" both solidified and changed. Hence, a self-narrative is not a reliable history of a person; memories, often invented and always embellished, are not a person.

Both my parents worked full-time in the grocery store they owned and had little time for childcare. My substitute parents were Grandma Rice and Murphy. Grandma Rice hugged me to her full bosom, told me that I was her wonderful little boy and sheltered me from a threatening world. Murphy lived in a small room in the basement, took his meals with the rest of us, and worked as an all-around handyman to keep the dilapidated grocery-store building from falling apart. I didn't know that Murphy liked his whiskey and was half in the bag most of the time. I just knew that he sang songs, joked about everything, and was fun to be around. When I wasn't with Grandma Rice, I was with Murphy.

When I was six years old, Grandma Rice and Murphy died within four months of each other. With the passing of my two principal caregivers, death clung to my shoulder like a pest I could not shake off. I hated and feared death, and as I grew older, death forced me to ask, "Who is this I that dies?"

Much later in life, I realized that "George Stanciu" is an artifact fashioned by culture and personal storytelling, devoid of substance and eternal permanence. Like every isolated, autonomous self, I believed that I was the center of existence to which everything should be ordered and sought to build up myself through the acquisition of knowledge, honor, and love. I laughed when I truly grasped that "George Stanciu" is an illusion, frail and fleeting, with no more permanency than a smoke ring, doomed to disappear into nothingness with the death of my body.

Memories Are Not Trustworthy

Our life narratives are based on what we think are accurate memories of past events, on the belief that memories are sealed within our skulls, immune from external influence. Memories not only fade, but they are also changed in many ways. The mere telling of an episode of our life narrative to others changes that memory; we enhance those memories that others respond to positively and downplay or edit out those memories that others dislike.

When a young mother shows to her child pictures on her cell phone of their trip to Disneyland and says, "Annie, you had such a great time talking to Mickey Mouse; that was the best part of your summer," she is implanting a memory in her child.[529]

Memories are not like a read-only computer file stored in the brain, and remembering is not like retrieving an uncorrupted digital document of our history accurately and permanently recorded. Daniel Offer and his colleagues at the Northwestern University Medical School examined "the differences between memories adults had of their teenage years and what they actually said when they were interviewed as adolescents" thirty-four years before. "The subjects' recollections were about the same as would have been expected by chance . . . the accuracy of recalled memories was uniformly poor."[530]

Recently, I received from a childhood friend a photograph of our third-grade class taken in 1944. I was shocked to see forty-two pupils, for I remembered the class as no more than eighteen students. I recollected that my classmates and I were from the solid middle class; so, I was surprised to see that most of the girls wore homemade dresses and scuffed shoes and that the boys had on well-worn clothes, except for two boys with ties, Joey Prinko and Patrick O'Neill, my two best friends, both of whom I had no difficulty recognizing. I immediately recalled that once when clowning around with them, I broke a bottle I held. I still have the scars on my index and middle fingers of my right hand. The memory of this foolish episode has been embellished from years of retelling to myself and others. Yet, the scars, like the childhood scars on my soul, confirm certain events happened.

Two summers ago, I visited a friend who lives in Provincetown, Massachusetts. When I returned to Santa Fe from Cape Cod, I had vivid, physical memories; I could feel my toes in beach sand, the glaring sun in my eyes, and the smell of sea breeze. The more I told these memories to myself and friends, the weaker the concrete memories became, until now they exist only in speech.[531] With the possible exceptions of wine connoisseurs, painters, and musicians, most of whom maintain that certain things are better left unsaid, verbalization erodes concrete experience until it is replaced by speech.

In summary, our life narratives are based on untrustworthy memories that are a shadow of concrete experience.

Who Am I? Redux

I was born in Pontiac, Michigan, and christened George Stanciu, but suppose at two months, I was adopted by the Li family, and they moved to Beijing, China, their home city. My new family named me Li Zhang Wei. That my given name Zhang Wei was second indicated I was first a member of the Li family or clan.

Zhang Wei's parents emphasized interdependence, group solidarity, social obligation, and personal humility. Zhang Wei was taught obedience, proper behavior, emotional restrain, and the value of group harmony. He understood himself in terms of his relation to a whole, to the Li family, to Chinese society, and, perhaps, to the Tao. "In Confucian human-centered philosophy man cannot exist alone," philosopher Hu Shih writes. "All action must be in the form of interaction between man and man."[532] Zhang Wei always saw himself as part of a larger whole. In the Chinese language, there is no word for "individualism;" the closest word is the one for "selfishness."[533]

If I were not adopted, my European-American parents often focused on my attributes, preferences, and judgments, making Georgie an individual. Parents in America aim to develop an autonomous self, and thus encourage independence, assertiveness, and self-expression in their offspring. Georgie was taught to "stick up for his rights and to fight his own battles."

In grade school, Georgie was trained in the ethos of capitalism. He competed with his fellow students to get the best grades and the most gold stars. In this way, Georgie learned *I succeed only if someone else fails*, and the converse — *if someone else succeeds, I must have failed*. Another lesson he learned was that *my success is entirely due to me, and no other person has a legitimate claim on its benefits* — a fundamental ethic of capitalism, where each person is responsible for his or her own success or failure.

By the end of the sixth grade, Georgie understood himself as an autonomous, isolated individual.

Both Li Zhang Wei and George Stanciu took their cultural-constructed selfs to be who they really were. While George took himself to be the center of the universe and Zhang Wei did not, both had natural self-love, although George's self-love was greatly enhanced by his individualistic culture.

That different cultures produce different "I"s is apparent in the twenty-first century. In America, the "I" is quick to anger; in the Etku Eskimo community, the "I" seldom experiences anger. The American "I" wishes for others to fail and becomes envious when they succeed, while the Lakota "I" takes pleasure in others' success. In America, the "I" is always lonely; in China, the "I" feels lonely only when separated from a lover or the family. To understand anything, the American "I" first looks to the smallest parts, while the Hopi "I" turns to the whole. The American "I" does not accept any man's word as proof of anything, while the Japanese "I" seeks guidance from masters. The

American "I" demands scientific demonstrations, whereas the Eastern Indian "I" wants a direct experience of the eternal.

In a roundabout way, I arrived at the central insight of the Buddha — the self is an illusion. In the Deer Park at Isiptana, the Buddha preached his second sermon, *The Discourse on Not-Self,* and "while this discourse was being spoken, the minds of the monks of the group of five were liberated from the taints by nonclinging."[534] Arguably, *anattā,* a Pāli word that literally means no-self, is the most important and most difficult concept in Buddhism, since it led the five monks to instant enlightenment, to *nirvana,* to "the annihilation of the illusion [of self], of the false idea of self."[535]

But if each one of us were merely a particular compound of body, sense perceptions, memories, and ideas, then no escape from *Samsara,* the never-ending wheel of birth and death, would be possible. The Buddha told his disciples, "There is, monks, an unborn, not become, not made, uncompounded . . . therefore an escape can be shown for what is born, has become, is made, is compounded."[536] I had no idea what the Buddha meant by the unborn, so I suspected that my answer to "Who am I?" missed an essential element of who I am.

In some mysterious way, I was more than my memories, which by themselves without storytelling were disconnected, and more than my self-narrative, whose main plot was my adolescent rebellion against all authority.

For me, of course, the unborn within me, my true self, was a complete mystery, so after stumbling around for years exploring Hinduism and Buddhism, I turned to the deepest understanding of the human person that Christianity offers.

As we saw in the preceding chapter, the Patristic Fathers embraced the theological insights of Pseudo-Dionysius the Areopogite: God is not any of the names used in the Hebrew Bible and the New Testament, not God of gods, Holy of holies, Cause of the ages, the still breeze, cloud, and rock.[537] God is *not* Mind, Greatness, Power, or Truth in any way we can understand, for He "cannot be understood, words cannot contain him, and no name can hold him. He is not one of the things that are, and he is no thing among things."[538] God is the Unnamable.

According to St. Gregory Palamas, we know the *energy* of God, not His *essence*: "Not a single created being has or can have any communion with or proximity to the sublime nature [of God]. Thus, if anyone has drawn near to God he evidently approached Him by means of His energy."[539] A person becomes close to God by participating in His energy, "by freely choosing to act well and to conduct [himself] with probity."[540] Here, Gregory distinguishes between God's essence, or substance (*ousia*), and His activity (*energeia*) in the world. The energy of God is experienced as Divine Light, such as the light of Mount Tabor or the light that blinded St. Paul on the Road to Damascus.

Given this understanding of God, the image of God within us means that the essence of each one of us is unnamable and that we are known to others only through our activity in the world, that is, through a socially-constructed self. We are unknowable to ourselves, although through meditation, or what the Patristic Fathers called contemplation, we can witness our thoughts, memories, and storytelling, and thus know that *we are not what we witness*. Through more advanced contemplation, we may experience Divine Light, the presence of God.

At the core of our being is the unnamable, the "empty mind" of Zen Buddhism, the "pure consciousness" of Hinduism, and the "spirit" of Christianity, although all words ultimately fail to capture our true self. We, the unenlightened, believe that the false self given to us by culture is permanent and fail to see that the false self is an illusion, with no more permanency than a smoke ring, destined to vanish with the death of the body. Because we take our culturally-given self for our true self, we fail to experience who we truly are. Our true self is always present, completely perfected with no need for development from us; we must merely step aside. Every spiritual master calls for the death of self and a spiritual rebirth beyond egoistic desires, beyond religious practices, beyond any given culture, beyond the dictates of society, into the law of love, into compassion for every living being.

The image of God within us also means that each one of us has the energy to transform the physical and social worlds we inhabit, either for good or evil. For instance, we are free to use the fruits of science for the benefit of life *or* the destruction of humanity, for creating polio vaccine *or* thermonuclear weapons, aids for life *or* instruments of death. Through such free choices, we either draw closer to God or more distant from Him.[541] Each one of us becomes what we choose.

Our freedom is virtually unlimited; we can thumb our nose at God, refuse to become who we truly are, and embrace a self of our own choosing. However, to freely abandon God, to exist in oneself, and to seek satisfaction in one's own being is not quite to become a nonentity but is to verge on non-being.[542] Hell is not the fiery pit of received Christianity, but the complete separation from God — forever. Heaven is not the reuniting with one's favorite dog or the blissful meeting with one's unknown relatives or the pleasure of conversing with the saints, not such "enthusiastic fantasies," but to "know more deeply the hidden presence by whose gift we truly live."[543] Heaven is joining the Holy Trinity in Love, as wonderfully expressed in the icon The Trinity, also called The Hospitality of Abraham, by Andrei Rublev. (See Figure 24.2.) The three angels in the icon are metaphors for the three persons of the Holy Trinity. The figures are arranged so that the lines of their bodies form a full circle. In motionless contemplation, each angel gazes into eternity. Because of inverse perspective, the focal point of the icon is in front of the painting on the viewer, inviting the viewer to complete the circle of angels, to join in a union with the Holy Trinity.

Figure 24.2. *The Trinity*, Andrei Rublev, 1427-1430. Courtesy of Wikimedia Commons.

Like every person, I live in two worlds, the temporal and the eternal. I love the taste of lamb curry, the sound of the cello, the fall foliage of New England, and wonder about the abundant beauty of Nature, where nothing is not beautiful, either to the eye or to the mind. Yet, this physical world, like "George Stanciu," is transient and eventually vanishes without leaving a trace.

I live among the rich and the poor, the powerful and the weak, the ambitious and the lazy, the good and the bad, the loving and the hateful. Grappling with death taught me how to live in this world. I now see that every person I meet in ordinary, daily affairs — the mailman, the bank teller, the butcher at Whole Foods, the obnoxious teenager down the street with his blaring boom box — is part human and part divine, a storytelling self, often confused, dislikable, and in pain, but always transient, and a mysterious self, deathless, an image of God, worthy of unconditional love.

Endnotes

1   A partial, conservative catalog of the political murders of the twentieth century is mind-boggling, unbelievable, but sadly undeniable. Deaths: World War I (military only): 9,700,000; Russian Revolution and Civil War: 9,000,000; forced collectivization: 3,000,000 Ukrainian peasants; Russian gulag: 1,000,000 political prisoners; Spanish Civil War: 1,200,000; World War II (military and civilian): 51,000,000; Nazi camps: 6,000,000 Jews and 6,000,000 Slavs, Gypsies, and political prisoners; Japanese Rape of Nanking: 300,000 Chinese; Allied bombing of Hamburg, Berlin, Cologne, and Dresden: 500,000 German civilians; Hiroshima and Nagasaki: 140,000 Japanese civilians; Vietnam War (military and civilian): 5,000,000; Chinese Great Leap Forward: 30,000,000. These numbers are low estimates. For the difficulty of estimating mass political murders, see Lewis M. Simons, "Genocide and the Science of Proof," *National Geographic Magazine* (January 2006): 28-35 and Timothy Snyder, "Holocaust: The Ignored Reality," *The New York Review of Books* (July 16, 2009).

2   Benedict XVI, reported by *L'Osservatore Romano* (July 2005).

3   Martin Heidegger, The *Spiegel* Interview (1966), Only a God Save Us. Found at http://www.archive.org/details/MartinHeidegger-DerSpiegelInterviewenglishTranslationonlyAGodCan

4   Friedrich Nietzsche, "On Truth and Lie in an Extra-Moral Sense," in *The Portable Nietzsche*, trans. Walter Kaufmann (New York: Vintage, 1979), p. 46.

5   Richard Rorty, *Consequences of Pragmatism: Essays, 1972-1980* (Minneapolis: University Of Minnesota Press, 1982), Introduction.

6   Giovanni Pico della Mirandola, *Oration on the Dignity of Man*. Found at http://www.archive.org/details/MartinHeidegger-DerSpiegel InterviewenglishTranslationonlyAGodCan.

7   Michel de Montaigne, "Apology for Raymond Sebond," in *The Complete Essays of Montaigne*, trans. Donald M. Frame (Stanford, CA: Stanford University Press, 1958), p. 330.

8   Ptolemy, *Almagest*, trans. G. J. Toomer (Princeton, NJ: Princeton University Press, 1998), pp. 35–36.

9   Aristotle, On the *Parts of Animals*. Found at http://classics.mit. edu/Aristotle/parts_animals.html

10   Aristotle, On the *Parts of Animals*, Bk. I, Ch. 1. Found at http://classics.mit.edu/Aristotle/parts_animals.html

11   Aristotle, *Metaphysics*, Bk. XII, Ch. 7.

12   Aristotle developed an extensive theory of ethics and politics, topics beyond the scope of this book. Here we present only a brief sketch of how Aristotle rooted ethics in nature. All plants and animals have their own *areté*, or excellence, represented by exemplars of the species sought for exhibits in botanical gardens, zoological parks, and biology textbooks.

To grasp the *areté* of *Homo sapiens*, Aristotle first notes that in nature the "whole is necessarily prior to the part . . . [and] that the polis exists by nature and that it is [therefore] prior to the individual." Consequently, "there is . . . an immanent impulse in all men towards" social association. Since "all things derive their character from their function and their capacity," the *areté* of a human being is to live well within society. "Man, when perfected, is the best of animals; but if he is isolated from law and justice, he is the worst of all." All quotations in this paragraph are from Aristotle, *The Politics of Aristotle*, trans. and ed. Ernest Barker (London: Oxford University Press, 1958).

13   Nicolaus Copernicus, *On the Revolutions of the Heavenly Spheres*, trans. R. Catesby Taliaferro in Great Books of the Western World, (Chicago: Encyclopedia Britannica, 1939), vol. 16, p. 508.

14   Nicolaus Copernicus, *On the Revolutions of the Heavenly Spheres*, trans. R. Catesby Taliaferro in Great Books of the Western World, (Chicago: Encyclopedia Britannica, 1939), vol. 16,, p. 506.

15   Nicolaus Copernicus, *On the Revolutions of the Heavenly Spheres*, trans. R. Catesby Taliaferro in Great Books of the Western World, (Chicago: Encyclopedia Britannica, 1939), vol. 16,, p. 525.

16   Galileo, "The Assayer," in *Discoveries and Opinions of Galileo* (New York: Doubleday Anchor, 1957), pp. 237-238.

17   Galileo, "The Assayer," in *Discoveries and Opinions of Galileo* (New York: Doubleday Anchor, 1957), p. 274.

18   Francis Bacon, *The New Organon and Related Writings* (Indianapolis, IN: Bobbs-Merrill, 1960 [1620]), p. 7.

19   Francis Bacon, *The New Organon and Related Writings* (Indianapolis, IN: Bobbs-Merrill, 1960 [1620]), p. 8.

20   Francis Bacon, *The New Organon and Related Writings* (Indianapolis, IN: Bobbs-Merrill, 1960 [1620]), pp. 3, 7. Italics added.

21   Francis Bacon, *The New Organon and Related Writings* (Indianapolis, IN: Bobbs-Merrill, 1960 [1620]); pp. 3, 7

22   Francis Bacon, *The New Organon and Related Writings* (Indianapolis, IN: Bobbs-Merrill, 1960 [1620]), p. 22.

23   Francis Bacon, *The New Organon and Related Writings* (Indianapolis, IN: Bobbs-Merrill, 1960 [1620]), p. 29.

24   Francis Bacon, *The New Organon and Related Writings* (Indianapolis, IN: Bobbs-Merrill, 1960 [1620]), p. 19.

25   Mircea Eliade, *Myths, Dreams, and Mysteries*, trans. Philip Mairet (New York: Harper, 1960), p. 43.

26   Mircea Eliade, *Myths, Dreams, and Mysteries*, trans. Philip Mairet (New York: Harper, 1960),

27   Francis Bacon, *The New Organon and Related Writings* (Indianapolis, IN: Bobbs-Merrill, 1960 [1620]), p. 103.

28   Francis Bacon, *The New Organon and Related Writings* (Indianapolis, IN: Bobbs-Merrill, 1960 [1620]),, p. 23.

29   René Descartes, *Discourse on Method* in *The Philosophical Works of Descartes*, trans. Elizabeth S. Haldane and G.R.T. Ross (Cambridge: Cambridge University Press, 1969), Vol. I, p. 115.

30   René Descartes, *Discourse on Method* in *The Philosophical Works of Descartes*, trans. Elizabeth S. Haldane and G.R.T. Ross (Cambridge: Cambridge University Press, 1969), Vol. I, p. 87.

31   René Descartes, *Discourse on Method* in *The Philosophical Works of Descartes*, trans. Elizabeth S. Haldane and G.R.T. Ross (Cambridge: Cambridge University Press, 1969), Vol. I, p. 92.

32   Isaac Newton, *Principia*, trans. Florian Cajori (Berkeley & Los Angeles: University of California Press, 1934), p. xviii.

33   See Alfred Korzybski, *Science and Sanity; an Introduction to Non-Aristotelian Systems and General Semantics* (Lakeville, CT: International Non-Aristotelian Library, 1958), Ch. IV and Supplement III. Found at http://esgs.free.fr/uk/art/sands.htm

34   See Ronald L. Numbers, "Science without God: Natural Laws and Christian Beliefs" in *When Science and Christianity Meet*, ed. David C. Lindberg and Ronald L. Numbers (Chicago: University of Chicago Press, 2008), p. 271.

35   Joshua Greene and Jonathan Cohen, "For the law, neuroscience changes nothing and everything," *Philosophical Transactions of the Royal Society London* B (2004) 359: 1780 - 1781. May be found at https://www.ncbi.nlm.nih.gov/pmc/articles/PMC1693457/

36   Joshua Greene and Jonathan Cohen, "For the law, neuroscience changes nothing and everything," *Philosophical Transactions of the Royal Society London* B (2004) 359: 1781. May be found at https://www.ncbi.nlm.nih.gov/pmc/articles/PMC1693457/

37   Joshua Greene and Jonathan Cohen, "For the law, neuroscience changes nothing and everything," *Philosophical Transactions of the Royal Society London* B (2004) 359: 1781. May be found at https://www.ncbi.nlm.nih.gov/pmc/articles/PMC1693457/

38   Joshua Greene and Jonathan Cohen, "For the law, neuroscience changes nothing and everything," *Philosophical Transactions of the Royal Society London* B (2004) 359: 1781. May be found at https://www.ncbi.nlm.nih.gov/pmc/articles/PMC1693457/

39   Francis Crick, *The Astonishing Hypothesis*, (New York: Scribner's, 1994), p. 3.

40   Joshua Greene and Jonathan Cohen, "For the law, neuroscience changes nothing and everything," *Philosophical Transactions of the Royal Society London* B (2004) 359: 1781. May be found at https://www.ncbi.nlm.nih.gov/pmc/articles/PMC1693457/

41   A similar argument was given by J.B.S. Haldane, a British geneticist and a committed Marxist. See J.B.S. Haldane, *Possible Worlds: And Other Essays* (London: Chatto and Windus, 1927; reprint, London: Transaction Publishers, 2002), p.209. Page reference is to the reprint edition.

42   Galileo, "The Assayer," in *Discoveries and Opinions of Galileo*, trans. Stillman Drake (Garden City, New York: Doubleday Anchor, 1957), p. 274.

43   Bertrand Russell, *Portraits from Memory* (New York: Simon & Schuster, 1956), p. 15.

44   Joshua Greene and Jonathan Cohen, "For the law, neuroscience changes nothing and everything," *Philosophical Transactions of the Royal Society London* B (2004) 359: 1781. May be found at https://www.ncbi.nlm.nih.gov/pmc/articles/PMC1693457/

45   Francis Crick, *The Astonishing Hypothesis*, (New York: Scribner's, 1994), p. 3.

46   Carl F. von Weizsacker, *The World View of Physics* (Chicago: University of Chicago Press, 1949), p. 203.

47   George F. R. Ellis, "Physics and the Real World," *Physics Today* **58** (July 2005): 53.

48   John Eccles, *Facing Reality: Philosophical Adventures by a Brain Scientist* (Berlin and New York: Springer-Verlag, 1970), p. 120.

49   Marvin Minsky, *The Society of Mind* (New York, Simon and Schuster, 1985), pp. 306, 307. Italics in the original.

50   Richard Dawkins, "Let's all stop beating Basil's car," found at https://www.edge.org/response-detail/11416.

51   Richard Dawkins, "Let's all stop beating Basil's car," found at https://www.edge.org/response-detail/11416.

52   See Ch. 8 Dumb Idea # 3: Materialism.

53   Stephen Hawking, quoted by David Deutsch, *The Fabric of Reality* (New York: Viking, 1997), pp. 177-178.

54   Steven Weinberg, *The First Three Minutes: A Modern View of the Origin of the Universe* (New York: Basic Books, 1977), p. 154.

55   See Richard Dawkins, *The Selfish Gene* (New York: Oxford University Press, 1976), p. 21.

56   Richard Dawkins, "Let's all stop beating Basil's car," found at https://www.edge.org/response-detail/11416.

57   William Provine, "Evolution and the Foundation of Ethics," *MBL Science* **3** (Winter 1988): 27.

58   Richard L. Rapson, *Denials of Doubt: An Interpretation of American History* (Lanham, MD: University Press of America, 1980), p. 5.

59   This definition applies to such varied groups as the Lakota Indians, Harvard University, and Google, Inc.

60   Eric Fromm, "The Creative Attitude," in *Creativity and Its Cultivation*, ed. Harold H. Anderson (New York: Harper & Row, 1959), p. 49.

61   S. M. Molema, *The Bantu: Past and Present* (Edinburgh: Green & Son, 1920), p. 136. Italics in the original.

62   Prince Modupe, *I Was a Savage* (New York: Harcourt, Brace, 1957), p. 110.

63   Alexis de Tocqueville, *Democracy in America*, trans. George Lawrence (New York: Harper & Row, 1966), p. 508.

64   Prince Modupe, *I Was a Savage* (New York: Harcourt, Brace, 1957), pp. 53-54.

65   Alexis de Tocqueville, *Democracy in America*, trans. George Lawrence (New York: Harper & Row, 1966), p. 429.

66   Alexis de Tocqueville, *Democracy in America*, trans. George Lawrence (New York: Harper & Row, 1966), p. 430.

67   Alexis de Tocqueville, *Democracy in America*, trans. George Lawrence (New York: Harper & Row, 1966), p. 429.

68   See Alexis de Tocqueville, *Democracy in America*, trans. George Lawrence (New York: Harper & Row, 1966), p. 429.

69   Alexis de Tocqueville, *Democracy in America*, trans. George Lawrence (New York: Harper & Row, 1966), p, 430.

70   Alexis de Tocqueville, *Democracy in America*, trans. George Lawrence (New York: Harper & Row, 1966), p. 431.

71   Alexis de Tocqueville, *Democracy in America*, trans. George Lawrence (New York: Harper & Row, 1966), p. 430.

72   Alexis de Tocqueville, *Democracy in America*, trans. George Lawrence (New York: Harper & Row, 1966), p. 429.

73   See Ch. 7 Things Exist Only in Relationship.

74   Anthony M. Kennedy, Sandra Day O'Connor, and David Hackett Souter, "Planned Parenthood of Southeastern Pennsylvania versus Casey," in *Constitutional Law: 1995 Supplement*, ed. Geoffrey R. Stone, et al. (Boston: Little, Brown, 1995), p. 955.

75   Nietzsche, "On Truth and Lie in an Extra-Moral Sense," p. 46.

76   George Levine et al., Speaking for the Humanities, *American Council of Learned Societies Occasional Paper*, no. 7 (1989): 9. May be found at http://archives.acls.org/op/7_Speaking_for_Humanities.htm

77   Friedrich Nietzsche, *The Will to Power*, trans. Walter Kaufmann and R. J. Hollingdale (New York: Vintage, 1968), p. 3.

78   David Foster Wallace, Kenyon College Commencement Address, 2005. May be found at https://www.wsj.com/articles/SB122178211966454607

79   All quotations in this paragraph are from Tocqueville, *Democracy in America*, p. 47.

80   Aristotle, *Nichomachean Ethics*, 1105a-b.

81   Plato, *Republic*, 557b

82   All quotations from David Brooks are from his article "Social Animal," *The New Yorker* (January 17, 2011).

83   Michael Ruse and Edward O. Wilson, "Evolution of Ethics," *New Scientist* **17** (October 1985): 51.

84   Francis Crick, *The Astonishing Hypothesis*, (New York: Scribner's, 1994), pp. 259, 7-8. (Italics in the original.)

85   Edward O. Wilson, *Consilience: The Unity of Knowledge* (New York: Knopf, 1998), p. 266.

86   P. B. Medawar and J. S. Medawar, *The Life Sciences: Current Ideas of Biology* (New York: Harper & Row, 1977), p. 165.

87   H. P. Yockey, "Information in Bits and Pieces," *BioEssays*, **17** no. 1 (1995): 85-88.

88   Steven Jay Gould, a leading evolutionary theorist, reached the same conclusion from a different argument, see Steven Jay Gould, "Humbled by the Genome's Mysteries," *The New York Times* (February 15, 2001).

89   John C. Polkinghorne, "The Nature of Physical Reality," *Zygon* **26** (June 1991): 221. Italics added.

90   R. C. Lewontin, *Biology as Ideology: The Doctrine of DNA* (New York: Harper, 1992), p. 10.

91   R. C. Lewontin, *Biology as Ideology: The Doctrine of DNA* (New York: Harper, 1992), p. 12.

92   Murray Gell-Mann, "Let's Call It Plectics," *Complexity* 1 (1995/1996), no. 5.

93   Francis Crick, *The Astonishing Hypothesis*, (New York: Scribner's, 1994), p. 3.

94   Alexis de Tocqueville, *Democracy in America*, trans. George Lawrence (New York: Harper & Row, 1966 [1835, 1840]), Appendix U, pp. 731-733. For narrative consistency, several verb tenses in the text have been changed to the past.

95   Alexis de Tocqueville, *Journey to America*, ed. J. P. Mayer, trans. George Lawrence (New York: Anchor: 1971), p. 360.

96   Alexis de Tocqueville, *Journey to America*, ed. J. P. Mayer, trans. George Lawrence (New York: Anchor: 1971), p. 11.

97   J. Hector St. John de Crèvecoeur, *Letters from an American Farmer*, ed. Susan Manning (New York: Oxford University Press, 1997), p. 126.

98   James Madison, "The Structure of the Government Must Furnish the Proper Checks and Balances between the Different Departments," Federalist No. 51.

99   John Locke, *Second Treatise of Government*, ed. C. B. Macpherson (New York: Hafner, 1980 [1690]), pp. 52, 66. May be found at http://www.gutenberg.org/dirs/7/3/7/7370/7370.txt.

100   Psalm 115:16. All Biblical quotations are from the RSV.

101   John Locke, *Second Treatise of Government*, ed. C. B. Macpherson (New York: Hafner, 1980 [1690]), p. 65. May be found at http://www.gutenberg.org/dirs/7/3/7/7370/7370.txt.

102   John Locke, *Second Treatise of Government*, ed. C. B. Macpherson (New York: Hafner, 1980 [1690]), 19. May be found at http://www.gutenberg.org/dirs/7/3/7/7370/7370.txt.

103   1 Tim. 6:17 All Biblical quotations are from the RSV.

104   John Locke, *Second Treatise of Government*, ed. C. B. Macpherson (New York: Hafner, 1980 [1690]), p. 21. May be found at http://www.gutenberg.org/dirs/7/3/7/7370/7370.txt.

105   John Locke, *Second Treatise of Government*, ed. C. B. Macpherson (New York: Hafner, 1980 [1690]), p. 28. May be found at http://www.gutenberg.org/dirs/7/3/7/7370/7370.txt.

106   John Locke, *Second Treatise of Government*, ed. C. B. Macpherson (New York: Hafner, 1980 [1690]), p 28. May be found at http://www.gutenberg.org/dirs/7/3/7/7370/7370.txt.

107   John Locke, *Second Treatise of Government*, ed. C. B. Macpherson (New York: Hafner, 1980 [1690]), p. 29. May be found at http://www.gutenberg.org/dirs/7/3/7/7370/7370.txt.

108   Giovanni Boccaccio, *The Decameron*, trans. Richard Aldington (New York: Dell, 1962), p. 33.

109   Thomas Gascoinge, quoted by William Manchester, *A World Lit Only By Fire* (Boston: Little, Brown, 1992), p. 131.

110   Martin Luther, "Open Letter to the Christian Nobility of the German Nation," in Martin Luther, *Three Treatises*, trans. Charles M. Jacobs (Philadelphia: Fortress Press, 1960), p. 14.

111   Eric Metaxas, *Martin Luther: The Man Who Rediscovered God and Changed the World* (New York: Viking, 2017), p. 1.

112   Robert A. Nisbet, *The Quest for Community* (New York: Oxford University Press, 1953), p. 90.

113   Perry Miller, "Individualism and the New England Tradition," in *The Responsibility of Mind in a Civilization of Machines: Essays by Perry Miller*, ed. John Crowell and Stanford J. Searl, Jr. (Amherst, MA: The University of Massachusetts Press, 1979.), p. 5, 6.

114   Alexis de Tocqueville, *The Old Régime and the French Revolution*, trans. Stuart Gilbert (New York: Doubleday, 1955 [1856]), p. 96.

115   Jacob Burckhardt, *The Civilization of the Renaissance in Italy*, trans. S. G. C. Middlemore (New York: Modern Library, 1954), p. 100.

116   Emmanuel Le Roy Ladurie, *Montaillou: The Promised Land of Error*, trans. Barbara Bray (New York: Vintage, 1979), p. 24.

117   Robert A. Nisbet, *The Quest for Community* (New York: Oxford University Press, 1953), p. 81.

118   Aristotle, *Politics*, 1253a.

119   Aristotle, *Politics*,, 1253a

120   Mircea Eliade, *The Sacred and the Profane: The Nature of Religion*, trans. Willard R. Trask (Orlando, FL: Harcourt, 1959), p. 178.

121   Aristotle, *Metaphysics* Bk. XII, Ch. 7.

122   Aquinas, *Summa Theologica*, I-I, Question 102, Article 1, Reply to Objection 3.

123   René Descartes, *Discourse on Method* (1637), in *The Philosophical Writings of Descartes Volume I*, trans. John Cottingham, Robert Stoothoff, and Dugald Murdoch (Cambridge: Cambridge University Press, 1985), Part II, p. 120.

124   David Bohm, *Wholeness and the Implicate Order* (London: Ark Paperbacks, 1983), p. 134. Italics in the original.

125   Richard P. Feynman, *The Feynman Lectures on Physics* (Reading, Mass.: Addison-Wesley, 1963), Ch. 46, p. 9.

126   See Ivo Kohler, *The Formation and Transformation of the Perceptual World*, trans. Harry Fiss (New York: International Universities Press, 1964).

127   See Richard Held, "Plasticity in Sensory-Motor Systems," *Scientific American* **213** (November 1965): 84-94.

128   Hazel Rose Markus and Shinobu Kitayama, "Culture and the self: Implications for cognition, emotion, and motivation," *Psychological Review* **98** (April 1991): 227.

129   See Sarah Lyall, "Spies Like Us: A Conversation with John le Carré and Ben Macintyre," *New York Times* (August 25, 2017).

130   Jules Henry, *Culture Against Man* (New York: Random House, 1963), pp. 295-296.

131   Alexis de Tocqueville, *Democracy in America*, trans. George Lawrence (New York: Harper & Row, 1966 [1835,1840]), p. 557.

132   National Poverty Center, *Extreme Poverty in the United States, 1996 to 2011*. May be found at http://npc.umich.edu/publications/policy_briefs/brief28/policybrief28.pdf

133   Martha Ross and Nicole Bateman, "Meet the Low-Wage Work Force," Brookings Institution (November 7, 2019). May be found at https://www.brookings.edu/research/meet-the-low-wage-workforce/

134   Jules Henry, *Culture Against Man* (New York: Random House, 1963), *Culture Against Man*, p. 296.

135   John Holt, *How Children Fail*, rev. ed. (Reading, MA.: Perseus, 1982), p. 274.

136   Karen Horney, *The Neurotic Personality of Our Time* (New York: Norton, 1937), p. 284.

137   Rollo May, *The Meaning of Anxiety*, rev. ed. (New York: Norton, 1977), p. 173.

138   See Robert Waldinger, "What makes a good life? Lessons from the longest study on happiness," Ted Talk (November 2015).

May be found at https://www.ted.com/talks/robert_waldinger_
what_makes_a_good_life_lessons_from_the_longest_study_on_
happiness?language=en

139   Robert Waldinger, "What makes a good life? Lessons from the
longest study on happiness." May be found at https://www.ted.com/
talks/robert_waldinger_what_makes_a_good_life_lessons_from_the_
longest_study_on_happiness?language=en

140   Robert Waldinger, "What makes a good life? Lessons from
the longest study on happiness," and George E. Vaillant, *Aging Well:
Surprising Guideposts to a Happier Life from the Landmark Study of Adult
Development* (New York: Little, Brown Spark, 2008). May be found at
https://www.ted.com/talks/robert_waldinger_what_makes_a_good_life_lessons_
from_the_longest_study_on_happiness?language=en

141   The writing on the postcard was a highly edited version of
Virginia Satire, *Self Esteem* (Millbrae, California: Celestial Arts, 1975).

142   Prince Modupe, *I Was a Savage* (New York: Harcourt, Brace,
1957), p. 110.

143   S. M. Molema, *The Bantu: Past and Present* (Edinburgh: Green
& Son, 1920), p. 136. Italics added.

144   For the history of Vaucanson's duck, see David F. Channell,
The Vital Machine: A Study of Technology and Organic Life (New York:
Oxford University Press, 1991), pp. 42-43; Etienne Benson, "Science
Historian Examines the 18th-Century Quest for 'Artificial Life'";
may be found at https://news.stanford.edu/news/2001/october24/
riskinprofile-1024.html and, Gaby Wood, "Living Dolls: A Magical
History Of The Quest For Mechanical Life." May be found at https://
www.theguardian.com/books/2002/feb/16/extract.gabywood

145   Francis Bacon, *The New Organon and Related Writings*
(Indianapolis, IN: Bobbs-Merrill, 1960 [1620]), p. 103.

146   Francis Bacon, *The New Organon and Related Writings*
(Indianapolis, IN: Bobbs-Merrill, 1960 [1620]), p. 18.

147   For instance, see Eric Kandel, *Psychiatry, Psychoanalysis, and
the New Biology of Mind* (Washington, D.C.: American Psychiatric
Publishing, 2005), p. 39.

148   Francis Crick, *The Astonishing Hypothesis*, (New York:
Scribner's, 1994), p. 3.

149   Joshua Greene and Jonathan Cohen, "For the law, neuroscience
changes nothing and everything," *Philosophical Transactions of the Royal
Society London* B (2004) 359: 1781. May be found at https://www.
ncbi.nlm.nih.gov/pmc/articles/PMC1693457/

150   Lynn Margulis and Dorion Sagan, "Strange Fruit on the Tree of Life," *The Sciences* **26** (May-June 1986): 43.

151   Richard Dawkins, *The Selfish Gene* (New York: Oxford University Press, 1976), p. 21.

152   Robert Cahn, "The 18 Arbitrary Parameters of The Standard Model In Your Everyday Life," *Review of Modern Physics* **68** (1996): 951.

153   Murray Gell-Mann, "Let's Call It Plectics," *Complexity* **1** (1995/1996), No. 5.

154   Stephen Hawking, quoted by David Deutsch, *The Fabric of Reality* (New York: Viking, 1997), pp. 177-178.

155   The quotation "the brain happens to be a meat machine" is widely attributed to Marvin Minsky, although I could not find the phrase "meat machine" in any article or book written by him.

156   Francis Crick, *The Astonishing Hypothesis*, (New York: Scribner's, 1994), p. 24.

157   Erwin Schrödinger, *What is Life? with Mind and Matter and Autobiographical Sketches* (Cambridge: Cambridge University Press, 1992), p. 153.

158   C. F. von Weizsäcker, *The History of Nature*, trans. Fred D. Wieck (Chicago: University of Chicago Press, 1949), pp. 14243.

159   Sir Charles Sherrington, *Man on His Nature*, (Cambridge: Cambridge University Press, 1963), p. 238.

160   Erwin Schrödinger, *What is Life? with Mind and Matter and Autobiographical Sketches* (Cambridge: Cambridge University Press, 1992) p. 164.

161   P. B. Medawar and J. S. Medawar, *The Life Sciences: Current Ideas of Biology* (New York: Harper & Row, 1977), p. 165.

162   P. W. Anderson, "More Is Different," *Science* **177** (4 August 1972): 393.

163   H. Allen Orr, "Awaiting a New Darwin," *The New York Review of Books*, **60**, No. 2 (February 7, 2013).

164   Recently, I discovered that Gottfried Wilhelm Leibniz, in 1714, gave an argument similar to "A Tour of Beethoven's Brain." He wrote, "Perception, and that which depends upon it, are inexplicable by mechanical causes, that is to say, by figures and motions. Supposing that there were a machine whose structure produced thought, sensation, and perception, we could conceive of it as increased in size with the same proportions until one was able to enter into its interior,

as he could into a mill. Now, on going into it he would find only pieces working upon one another, but never would he find anything to explain perception." See Gottfried Wilhelm Leibniz, *The Monadology* in *Discourse on Metaphysics and The Monadology*, trans. George R. Montgomery (Buffalo, NY: Prometheus Books, 1992), p. 70.

165   Oliver Sacks, *The Man Who Mistook His Wife for a Hat* (New York: Summit, 1987), p. 39.

166   Antonio Damasio, *The Feeling of What Happens: Body and Emotions in the Making of Consciousness* (New York: Harcourt Brace, 1999), p. 9.

167   Christof Koch, quoted by Julie Wakefield, "A Mind for Consciousness," *Scientific American* **285** (July 2001): 37.

168   Francis Crick and Christof Koch, "What Are the Neural Correlates of Consciousness?" in *23 Problems in Systems Neuroscience*, ed. J. Leo van Hemmen and Terrence J. Sejnowski (New York: Oxford University Press, 2005), p. 474.

169   V. S. Ramachandran, *The Tell-Tale Brain: A Neuroscientist's Quest for What Makes Us Human* (New York: Norton, 2011), p. 248.

170   V. S. Ramachandran, *The Tell-Tale Brain: A Neuroscientist's Quest for What Makes Us Human* (New York: Norton, 2011), p. 249. Ramachandran's definition of qualia is on page 248 and is incorporated in this quotation.

171   Orr, "Awaiting a New Darwin," *The New York Review of Books*, **60**, No. 2 (February 7, 2013). Italics in the original.

172   Roger Sperry, "Mind, Brain and Humanist Values," in *New Views of the Nature of Man*, ed. John R. Platt (Chicago: University of Chicago Press, 1965), pp. 78, 82.

173   R. W. Sperry, "Search for Beliefs to Live by Consistent with Science," *Zygon* **26** (June 1991): 246.

174   R. W. Sperry, "Psychology's Mentalist Paradigm and the Religion/Science Tension," *American Psychologist* **43** (August 1988): 607, and "Search for Beliefs to Live by Consistent with Science," *Zygon* **26** (June 1991): 238.

175   Erwin Schrödinger, *Nature and the Greeks and Science and Humanism* (Cambridge: Cambridge University Press, 1996), p. 95.

176   J. Robert Oppenheimer, quoted by Richard Rhodes, *Dark Sun: The Making of the Hydrogen Bomb* (New York: Simon & Schuster, 1996), p. 578.

177	Steven Weinberg, "The Trouble with Quantum Mechanics," *The New York Review of Books* **64** (January 19, 2017).

178	For the strange history of this quotation attributed to Lord Kelvin, see Limits of Classical Physics in the Wikipedia article on William Thomas, 1st Baron Kelvin. May be found at http://en.wikipedia.org/wiki/Lord_Kelvin#Limits_of_classical_physics

179	Albert Michelson, quoted in Michelson, Albert Abraham (1852-1931). May be found at http://scienceworld.wolfram.com/biography/Michelson.html

180	Philipp von Jolly, quoted in "Philipp von Jolly," Wikipedia.

181	Ernest Rutherford, quoted by David C. Cassidy, Gerald James Holton, Gerald Holton, and Floyd James Rutherford, *Understanding Physics* (New York: Springer, 2002), p. 632.

182	Richard P. Feynman, *The Feynman Lectures on Physics Vol. III* (Reading, Mass.: Addison-Wesley, 1963), Ch. 1, p. 1. Italics in the original.

183	Richard P. Feynman, *The Feynman Lectures on Physics Vol. III* (Reading, Mass.: Addison-Wesley, 1963), Ch. 1, p. 1. Italics in the original., Ch. 1, p. 5.

184	May be found at https://www.hitachi.com/rd/research/materials/quantum/doubleslit/index.html

185	Richard P. Feynman, *The Feynman Lectures on Physics Vol. III* (Reading, Mass.: Addison-Wesley, 1963), Ch. 1, p. 1. Italics in the original., Ch. 1, p. 10.

186	See Werner Heisenberg, *Physics and Philosophy: The Revolution in Modern Science* (New York: Harper & Row, 1958), p. 54.

187	*Physics and Philosophy: The Revolution in Modern Science* (New York: Harper & Row, 1958, p. 186.

188	Werner Heisenberg, "The Representation of Nature in Contemporary Physics," *Daedalus* **87** (no. 3): 99-100.

189	David Bohm, *Wholeness and the Implicate Order* (London: Ark Paperbacks, 1983), p. 134. Italics in the original.

190	Galileo, "The Assayer," in *Discoveries and Opinions of Galileo*, trans. Stillman Drake (Garden City, New York: Doubleday Anchor, 1957), p. 274.

191	Francis Bacon, *The New Organon: Or the True Directions Concerning the Interpretation of Nature* (Indianapolis, IN: Bobbs-Merrill, 1960 [1620]), p. 22.

192   See H. Allen Orr, "Awaiting a New Darwin," *The New York Review of Books*, **60** (February 7, 2013).

193   Neils Henrik Bohr, *Essays 1958-1962 on Atomic Physics and Human Knowledge* (New York: Wiley, 1963), p. 15.

194   Eugene P. Wigner, *Symmetries and Reflections: Scientific Essays of Eugene P. Wigner* (Woodbridge, Conn.: Ox Bow Press, 1979), p. 172.

195   Max Born, *Physics in My Generation* (London & New York: Pergamon, 1956), p. 48.

196   Freeman Dyson, *Disturbing the Universe* (New York: Harper & Row, 1979), p. 249.

197   Maximilian Schlosshauer, Johannes Kofler, and Anton Zeilinger, "A Snapshot of Foundational Attitudes Toward Quantum Mechanics." May be found at https://arxiv.org/pdf/1301.1069v1.pdf

198   To whom this phrase should be attributed to, see N. David Mermin, "Could Feynman Have Said This?" Physics Today **57**, 5, 10 (2004). May be found at https://physicstoday.scitation.org/doi/full/10.1063/1.1768652

199   Steven Weinberg, "The Trouble with Quantum Mechanics.", *The New York Review of Books* **64** (January 19, 2017).

200   Steven Weinberg, "The Trouble with Quantum Mechanics.", *The New York Review of Books* **64** (January 19, 2017).

201   See Chapter 2 Dumb Idea #1: Determinism.

202   See Chapter 9 Dumb Idea # 3: Materialism.

203   Francis Crick, *The Astonishing Hypothesis*, (New York: Scribner's, 1994), p. 3.

204   Francis Crick, *The Astonishing Hypothesis*, (New York: Scribner's, 1994), p. 3

205   See Richard Dawkins, *The Selfish Gene* (New York: Oxford University Press, 1976), p. 21.

206   Sigmund Freud, *Civilization and Its Discontents*, trans. and ed. James Strachey (New York: Norton, 1961), pp. 24-25.

207   Steven Weinberg, *The First Three Minutes* (New York: Basic Books, 1977), p. 154.

208   Stephen Hawking, quoted by David Deutsch, *The Fabric of Reality* (New York: Viking, 1997), pp. 177-178.

209   Roger Sperry, "Mind, Brain and Humanist Values," in *New Views of the Nature of Man*, ed. John R. Platt (Chicago: University of Chicago Press, 1965), pp. 78, 82.

210   John A. Wheeler, "Genesis and Observership," in *Foundational Problems in the Special Sciences,* ed. Robert E. Butts and Jaakko Hintikka (Dordrecht, Holland: Reidel, 1977), p. 18.

211   See Chapter 15 A Frog Tells Me Who I Am.

212   Erwin Schrödinger, *Mind and Matter* (Cambridge: Cambridge University Press, 1959), p. 2.

213   Francis Bacon, *The New Organon and Related Writings* (Indianapolis, IN: Bobbs-Merrill, 1960 [1620]), p. 29.

214   Bacon, *The New Organon*, p. 103.

215   The data on train travel in nineteenth-century Europe is from Orlando Figes, *The Europeans: Three Lives and the Making of a Cosmopolitan Culture* (New York: Metropolitan Books, 2019).

216   Heinrich Heine, quoted by Wolfgang Schivelbusch, *The Railway Journey: The Industrialization of Time and Space in the Nineteenth Century* (Berkeley, CA: University of California Press, 2014), p. 37.

217   Charles Baudelaire, *The Painter of Modern Life and Other Essays*, trans. and ed. John Mayne (New York: Phaidon, 1965), p. 13.

218   Karl Marx, *Grundrisse: Foundations of the Critique of Political Economy*, written 1857–61, published in German 1939–41, trans. Martin Nicolaus, p. 449. May be found at https://www.marxists.org/archive/marx/works/1857/grundrisse/index.htm

219   See Barry Levinson's film *Avalon* for a brilliant rendering of how the automobile killed the extended family in America.

220   Adam Smith, *The Wealth of Nations* [1776]), Bk. I, Ch. I. May be found at http://www.econlib.org/library/Smith/smWN.html

221   Adam Smith, *The Wealth of Nations* [1776]), Bk. I, Ch. I. May be found at http://www.econlib.org/library/Smith/smWN.html

222   Adam Smith, *The Wealth of Nations* [1776]), May be found at http://www.econlib.org/library/Smith/smWN.html, Bk. V, Ch. I.

223   Alexis de Tocqueville, *Democracy in America*, trans. George Lawrence (New York: Harper & Row, 1966 [1835, 1840]), p. 556.

224   Adam Smith, *The Wealth of Nations* [1776]) May be found at http://www.econlib.org/library/Smith/smWN.html, Bk. V, Ch. I.

225   Steven Johnson, *Extra Life: A Short History of Living Longer* (New York: Riverhead Books, 2021).

226   Neils Henrik Bohr, *Essays 1958-1962 on Atomic Physics and Human Knowledge* (New York: Wiley, 1963), p. 15.

227   "Penetration rate of smartphones in selected countries 2020," May be found at https://www.statista.com/statistics/539395/smartphone-penetration-worldwide-by-country/

228   May be found at https://www.youtube.com/watch?v=4wH878t78bw.

229   See "Hedge fund forced Cabela's merger, decimated jobs in Sidney, Nebraska," May be found at https://www.youtube.com/watch?v=UatnTSwEUoc

230   Sophocles, *Antigone*, trans. R. C. Jebb, line 614. May be found at http://classics.mit.edu/Sophocles/antigone.html

231   John Gramlich, "5 facts about Americans and Facebook." May be found at http://www.pewresearch.org/fact-tank/2018/04/10/5-facts-about-americans-and-facebook/

232   Aaron Smith, "What people like and dislike about Facebook." May be found at http://www.pewresearch.org/fact-tank/2014/02/03/what-people-like-dislike-about-facebook/

233   Geoffrey O'Brien, *The Phantom Empire: Movies in the Mind of the 20th Century* (New York: Norton, 1993), p. 218.

234   *The Nielson Total Audience Report, Q1, 2016.* May be found at https://www.nielsen.com/us/en/insights/report/2016/the-total-audience-report-q1-2016/

235   Clay Shirky, "How Social Media Can Make History, "Ted Talk, June 2009. May be found at https://www.ted.com/talks/clay_shirky_how_cellphones_twitter_facebook_can_make_history

236   "Andy Chan, Artificial Intelligence and the Future of Work, "Ted Talk, March 2019. May be found at https://www.ted.com/talks/andy_chan_artificial_intelligence_and_the_future_of_work

237   Andy Chan, "Artificial Intelligence and the Future of Work," Ted Talk, March 2019. May be found at https://www.ted.com/talks/andy_chan_artificial_intelligence_and_the_future_of_work

238   Adam Smith, *The Wealth of Nations* [1776])May be found at http://www.econlib.org/library/Smith/smWN.html, Bk. V, Ch. I.

239   Mircea Eliade, *Myths, Dreams, and Mysteries*, trans. Philip Mairet (New York: Harper, 1960), p. 43.

240   The Spirit of the Times," *The United States Democratic Review* **33** (Issue 9, Sept 1853): 261. May be found at https://babel.hathitrust.org/cgi/pt?id=mdp.39015013470599&view=1up&seq=271

241   Alexis de Tocqueville, *Democracy in America*, trans. George Lawrence (New York: Harper & Row, 1966), p. 536.

242   See Robert E. Lane, *The Loss of Happiness in Market Democracies* (New Haven, CN: Yale University Press, 2001), p. 5.

243   Daniel Kahneman and Angus Deaton, "High income improves evaluation of life but not emotional well-being," *Proceedings of the National Academy of Sciences of the United States of America* (August 4, 2010) 107 (**38**): 16489–16493. May be found at http://www.pnas.org/content/107/38/16489.full

244   The Microsoft Bing Dictionary is out of print and is no longer available online.

245   Genesis 2:19-20. RSV

246   Francis Bacon, *The New Organon and Related Writings* (Indianapolis, IN: Bobbs-Merrill, 1960 [1620]), pp. 3, 7. Italics added.

247   Francis Bacon, *The New Organon and Related Writings* (Indianapolis, IN: Bobbs-Merrill, 1960 [1620])., p. 29.

248   Francis Bacon, *The New Organon and Related Writings* (Indianapolis, IN: Bobbs-Merrill, 1960 [1620]), p. 15.

249   Francis Bacon, *The New Organon and Related Writings* (Indianapolis, IN: Bobbs-Merrill, 1960 [1620])

250   Francis Bacon, *The New Organon and Related Writings* (Indianapolis, IN: Bobbs-Merrill, 1960 [1620]), p. 103.

251   Francis Bacon, *The New Organon and Related Writings* (Indianapolis, IN: Bobbs-Merrill, 1960 [1620]), p. 23.

252   H. Allen Orr, "Awaiting a New Darwin," *The New York Review of Books*, **60**, No. 2 (February 7, 2013).

253   Francis Bacon, "Of Atheism," in *Essays, Civil and Moral.* May be found at http://www.bartleby.com/3/1/

254   Francis Bacon, "Of Atheism," in *Essays, Civil and Moral.* May be found at http://www.bartleby.com/3/1/.

255   Karl Marx, "Economic and Philosophical Manuscripts," in *Early Writings*, trans. Rodney Livingstone and Gregor Benton (New York: Penguin, 1992), p. 356. Italics in the original.

256   Karl Marx, "Economic and Philosophical Manuscripts," in *Early Writings*, trans. Rodney Livingstone and Gregor Benton (New York: Penguin, 1992), p. 356. Italics in the original., p. 357.

257   The numbers in the text do not include the 100,000 Cambodian peasants killed in the 1972-1973 bombing campaign initiated by President Richard Nixon and Secretary of State Henry Kissinger. The number of bombs dropped on Cambodia was over three times that dropped on Japan in World War II. So much for technological progress.

258   Aleksandr I. Solzhenitsyn, *The Gulag Archipelago, 1918-1956: An Experiment in Literary Investigation Part III and IV*, trans. Thomas P. Whitney (New York: Harper & Row, 1973), pp. 54, 126.

259   Dmitri Vitkovsky, quoted in Aleksandr I. Solzhenitsyn, *The Gulag Archipelago, 1918-1956: An Experiment in Literary Investigation Part III and IV*, trans. Thomas P. Whitney (New York: Harper & Row, 1973), p. 99.

260   "Excess mortality in the Soviet Union under Joseph Stalin." May be found at https://en.wikipedia.org/wiki/Excess_mortality_in_the_Soviet_Union_under_Joseph_Stalin

261   Primo Levi, *The Drowned and the Saved*, trans. Raymond Rosenthal (New York: Vintage, 1989), p. 179.

262   Adolf Hitler, *Mein Kampf*, 1925 and 1926, trans. James Murphy, Vol. II, Ch. 15. May be found at http://www.greatwar.nl/books/meinkampf/meinkampf.pdf

263   Adolf Hitler, *Mein Kampf*, 1925 and 1926, trans. James Murphy, Vol. I, Ch. 4, Ch. 4, Ch. 11, Ch. 11. May be found at http://www.greatwar.nl/books/meinkampf/meinkampf.pdf

264   The Dutchman Nikko Tinbergen, an outspoken foe of Nazism, spent two years interned in a Nazi concentration camp.

265   Konrad Lorenz, quoted by Benno Muller-Hill, *Murderous Science: Elimination by Scientific Selection of Jews, Gypsies, and Others in Germany, 1933-1945* (Cold Spring Harbor, New York: Cold Spring Harbor Laboratory Press, 1997), p. 15.

266   Konrad Lorenz, quoted by Robert Jay Lifton, *The Nazi Doctors: Medical Killing and the Psychology of Genocide* (New York: Basic Books, 2000), p. 134.

267   *Herman Rauschning, Hitler Speaks: A series of Political Conversations with Adolf Hitler on His Real Aims* (Whitefish, Montana: Kessinger, 2010 [1939]), p. 220.

268   See Norman Owen-Smith, "Territoriality in the White Rhinoceros (Ceratotherium simum) Burchell," Nature **231** (4 June 1971): 295.

269   Konrad Z. Lorenz, *King Solomon's Ring*, trans. Marjorie Kerr Wilson (New York: Meridian, 1997 [1952]), p. 155.

270   Keller Breland and Marian Breland, "The Misbehavior of Organisms," *American Psychologist* **16** (1961): 681-684.

271   Sophocles, *Antigone*, trans. R. C. Jebb, line 614. May be found at http://classics.mit.edu/Sophocles/antigone.html.

272   Helen Keller, *The Story of My Life* (Mineola, New York: Dover, 1996 [1903]).

273   Helen Keller, *The World I Live In* (New York: The Century Co., 1904, 1908).

274   Helen Keller, *The World I Live In* (New York: The Century Co., 1904, 1908).

275   Robert Fantz, "The Origin of Form Perception," *Scientific American* **204** (May 1961):69.

276   Daniel G. Freedman, *Human Infancy: An Evolutionary Perspective* (Hillsdale, New Jersey: Erlbaum, 1974), p. 30.

277   James L. Gould, *Ethology: Mechanisms and Evolution of Behavior* (New York: Norton, 1982), p. 264.

278   Confucius, *The Humanist Way in Ancient China: Essential Works of Confucianism*, ed. and trans. Ch'u Chai and Winberg Chai (New York: Bantam Books, 1965), p. 44.

279   Arthur Eddington, *The Philosophy of Physical Science* (Ann Arbor: University of Michigan Press, 1958), p. 160.

280   Standing Bear, *Land of the Spotted Eagle (*Lincoln: University of Nebraska Press, 1978*)*, pp. 69-70.

281   John T. Prince, *Courses and Methods: A Handbook for Teachers* (Boston: Ginn, 1892), p. 188.

282   Standing Bear, *Land of the Spotted Eagle (Lincoln: University of Nebraska Press, 1978)*, p. 69.

283   Charles Darwin, *The Voyage of the Beagle* (New York: Dutton, 1967), p. 8.

284   Mozart, quoted by Joseph Machlis, *The Enjoyment of Music* (New York: Norton, 1963), p. 308.

285   Edward Holmes, *The Life of Mozart* (London: Dent, 1845), p. 251.

286   Abraham Maslow, "Creativity in Self Actualizing People," in *Creativity and Its Cultivation* (New York: Harper & Row, 1959), p. 85.

287   S. M. Ulam, *Adventures of a Mathematician* (New York: Scribner's, 1976), p. 163

288   Albert Einstein, "Autobiographical Notes," in *Albert Einstein: Philosopher-Scientist*, ed. Paul Schilpp (New York: Harper & Row, 1959), p. 33.

289   C.P. Snow, "The Moral Un-neutrality of Science," in his *Public Affairs* (New York: Scribner's, 1971), p. 189.

290   John Donne, *Essays*, Meditation XVII.

291   See Thomas Aquinas, *Summa Theologica* Part 1, Question 78, Reply to objection 5.

292   Donald R. Griffin, "Animal Thinking," *American Scientist* **72** (September-October 1984): 456.

293   René Descartes, Letter to Henry More, February 5, 1649, quoted by Mirko D. Grmek, "A Survey of the Mechanical Interpretations of Life from the Greek Atomists to the Followers of Descartes," in *Biology, History and Natural Philosophy*, ed. Allen Breck and Wolfgang Yourgau (New York: Plenum, 1972), Note 8, Ch. 2. Italics added.

294   René Descartes, *Treatise on Man*, trans. Thomas S. Hall (Cambridge: Harvard University Press, 1972), pp. 21, 28-29.

295   René Descartes, *Treatise on Man*, trans. Thomas S. Hall (Cambridge: Harvard University Press, 1972), pp. 71, 21.

296   René Descartes, *Description of the Human Body* in *The Philosophical Writings of Descartes: Volume 1*, trans. John Cottingham (Cambridge: Cambridge University Press, 1985), p. 315.

297   Descartes, *Treatise on Man*, pp. 36-37.

298   Gary Francione, *Introduction to Animal Rights: Your Child or the Dog?* (Philadelphia, PA: Temple University Press, 1999).

299   Thomas H. Huxley, *Method and Results* (New York & London: 1925), pp. 216, 218.

300   Thomas H. Huxley, *Method and Results* (New York & London: 1925), p. 240.

301   Thomas H. Huxley, *Method and Results* (New York & London: 1925), p. 240.

302   Alexis de Tocqueville, *Democracy in America*, trans. George Lawrence (New York: Harper & Row, 1966 [1835, 1840]), p. 516.

303   Edward O. Wilson, *On Human Nature* (Cambridge: Harvard University Press, 1978), p. 16.

304   Kenneth Oakley, *Antiquity* **30** (1956): 4.

305   Jane Goodall, *In the Shadow of Man* (Boston: Houghton, Mifflin, Harcourt: 1971), p. 36. Italics in original.

306   Jane Goodall, *In the Shadow of Man* (Boston: Houghton, Mifflin, Harcourt: 1971), p. 37.

307   Louise Leakey, quoted by Jane Goodall and Phillip Berman, *Reason for Hope: A Spiritual Journey* (New York: Grand Central, 1999), p. 67.

308   S. Tebbich, M. Taborsky, B. Fessl, and D. Blomqvist, "Do woodpecker finches acquire tool-use by social learning?" *Proceedings of the Royal Society London* B (2001) **268**: 2189-2193.

309   S. Tebbich, M. Taborsky, B. Fessl, and D. Blomqvist, "Do woodpecker finches acquire tool-use by social learning?" *Proceedings of the Royal Society London* B (2001) **268**: p. 2192.

310   Jane Goodall, *In the Shadow of Man* (Boston: Houghton, Mifflin, Harcourt: 1971), p. 236.

311   Kate Wong, "Tiny Genetic Differences between Humans and Other Primates Pervade the Genome," *Scientific American* 311 (September 1, 2014).

312   Charles Darwin, *The Expression of the Emotions in Man and Animals* (London: John Murray, 1872), p. 51.

313   Charles Darwin, *The Expression of the Emotions in Man and Animals* (London: John Murray, 1872), p. 250.

314   Carroll E. Izard, *Human Emotions* (New York: Plenum, 1977), p. 6.

315   Jane Goodall, *In the Shadow of Man* (Boston: Houghton, Mifflin, Harcourt: 1971), p. 237.

316   Michael Seres, quoted by Frans B. M. de Waal, *Good Natured: The Origins of Right and Wrong in Humans and Other Animals* (Cambridge, MA: Harvard University Press, 1997), pp. 59-60.

317   Frans de Waal, *The Ape and the Sushi Master: Cultural Reflections of a Primatologist* (New York: Basic Books, 2001), p. 327.

318   See the documentary *Weapons of the Spirit* (1987), director Pierre Sauvage; DVD available from http://www.chambon.org/weapons_en.htm.

319   David P. Watts and John C. Mitani, "Hunting and Prey Switching by Chimpanzees (*Pan troglodytes schweinfurthii*) at Ngogo," *International Journal of Primatology* **36** No. 4 (July 19, 2015).

320   David P. Watts, quoted by Michael Marshall, "Chimpanzees over-hunt monkey prey almost to extinction," *BBC Earth* (**28** July 2015).

321   Frans de Waal, *Our Inner Ape: A Leading Primatologist Explains Why We Are Who We Are* (New York: Riverhead Books, 2006), p. 28.

322   Jane Goodall, *Beyond Innocence: An Autobiography in Letters: The Later Years* (New York: Mariner Books, 2002), p. 207.

323   Frans de Waal, *Our Inner Ape: A Leading Primatologist Explains Why We Are Who We Are* (New York: Riverhead Books, 2006), p. 142.

324   For a discussion of genocide in the century of mass murder, see Lewis M. Simons, "Genocide and the Science of Proof," *National Geographic Magazine* (January 2006): 28-35 and Timothy Snyder, "Holocaust: The Ignored Reality," *The New York Review of Books* (July 16, 2009).

325   Robert Sapolsky, *Behave: The Biology of Humans at Our Best and Worst* (New York: Penguin Press, 2017), p. 648.

326   Robert Sapolsky, *Behave: The Biology of Humans at Our Best and Worst* (New York: Penguin Press, 2017), p. 651.

327   R. Allen Gardner and Beatrice T. Gardner, "Teaching Sign Language to a Chimpanzee," *Science* **165** (15 August 1969): 664-672.

328   Roger S. Fouts, Deborah H. Fouts, and Thomas E. Van Cantfort, "The Infant Loulis Learns Signs from Cross-Fostered Chimpanzees," in *Teaching Sign Language to Chimpanzees*, ed. R. Allen Gardner, Beatrix T. Gardner, and Thomas E. Van Cantfort (Albany, NY: SUNY Press, 1989), p. 281.

329   Francine G. Patterson, "Linguistic Capabilities of a Lowland Gorilla," in *Language Intervention from Ape to Child*, ed. Richard L. Schiefelbush and John H. Hollis (Baltimore: University Park Press, 1979), pp. 325-356.

330   See the documentary *Project Nim*, director James Marsh, (BBC Films, 2011).

331   See Herbert S. Terrace, *Nim* (New York: Knopf, 1979) and also the documentary *Project Nim*.

332   Herbert S. Terrace, L. A. Petitto, R. J. Sanders, and T. G. Bever, "Can an Ape Create a Sentence?" *Science* **206** (23 November 1979): 900.

333   Herbert S. Terrace, L. A. Petitto, R. J. Sanders, and T. G. Bever, "Can an Ape Create a Sentence?" *Science* **206** (23 November 1979): p. 891.

334   Herbert S. Terrace, L. A. Petitto, R. J. Sanders, and T. G. Bever, "Can an Ape Create a Sentence?" *Science* **206** (23 November 1979): pp. 895-896.

335   Thomas A. Sebeok and Jean Umiker-Sebeok, "Performing Animals: Secrets of the Trade," *Psychology Today* **13** (November 1979): 91.

336   Noam Chomsky, quoted in *Time* (10 March 1980), p. 57.

337   For a witty, short history of the efforts to teach American Sign Language to nonhuman primates, view the last twenty-five minutes of Robert Sapolsky, *Human Behavioral Biology*, Lecture 23 On Language.

338   For a witty, short history of the efforts to teach American Sign Language to nonhuman primates, view the last twenty-five minutes of Robert Sapolsky, *Human Behavioral Biology*, Lecture 23 On Language. Human speech also requires the correct anatomy, see Philip Lieberman, "Why Human Speech Is Special," *TheScientist* (July 1, 2018).

339   Alexis de Tocqueville, *Democracy in America*, trans. George Lawrence (New York: Harper & Row, 1966), p. 516.

340   J. Y. Lettvin, H. R. Maturana, W. S. McCulloch, and W. H. Pitts, "What the Frog's Eye Tells the Frog's Brain," *Proceedings of the Institute of Radio Engineers* **47** (November 1959): 1940.

341   Figure 15.1 is a composite of Shutterstock images: Michiel de Wit, "Northern Leopard Frog (*Lithobates pipiens*);" Africa Studio, "Fishhook with worm isolated on white background;" Irink, "Bee isolated on white;" Anneka, "Mealworm or worm on a fishing hook as bait;" and Vnlit, "Dragonfly macro isolated on white background."

342   Jacob von Uexküll, quoted by Josef Pieper, *Leisure: The Basis Culture*, trans. Alexander Dru (New York: Mentor, 1963), p. 86.

343   Konrad Lorenz, *On Aggression* (New York: Harcourt & World, 1963), pp. 117-118.

344   Konrad Lorenz, *King Solomon's Ring* (New York: Crowell, 1952), p. 140.

345   Wolfgang Kohler, *The Mentality of Apes*, trans. Ella Winter (New York: Harcourt, Brace, 1931), pp. 320-321.

346   Francine G. Patterson, "Conversations with a Gorilla," *National Geographic* **154** (October 1978): 456, 459.

347   Uexküll, quoted by Josef Pieper, *Leisure: The Basis Culture*, p. 85.

348   Oliver Sacks, *The Man Who Mistook His Wife for a Hat and Other Clinical Tales* (New York: 1998), p. 9.

349   E.O. Wilson, see Bert Hölldobler and Edward O. Wilson, *Journey to the Ants: A Story of Scientific Exploration* (Harvard University Press, 1994), p. 68.

350   Henri Poincaré, *The Value of Science* (New York: Dover, 1958), p. 8.

351   Aristotle, *De Anima*, Bk. III, Ch. 8, 431b.

352   *Chandogya Upanishad*, 6.12-14. Found at https://www.hinduwebsite.com/sacredscripts/hinduism/upanishads/chandogya.asp

353   Thomas Aquinas, quoted by *Josef Pieper, Leisure: The Basis Culture*, p. 88.

354   Gershom Scholem, *Kabbalah* (Jerusalem: Keter, 1974), p. 152.

355   Jalaluddin Rumi, *Signs of the Unseen: The Discourses of Jalaluddin Rumi*, trans. W. M. Thackston, Jr. (Putney, Vermont: Threshold Books, 1994), p. 59.

356   Lao Tzu, *Tao Te Ching*, No. 16.

357   Billy Yellow, interview by David Maybury-Lewis, *Millennium*, aired on PBS, 1992.

358   Jean Biggs, *Never in Anger: Portrait of an Eskimo Family* (Cambridge: Harvard University Press, 1970), p. 74.

359   Ashley Montagu, interview Dennis Wholey, *Discovering Happiness: Personal Conversations About Getting the Most Out of Life*, (New York: Avon, 1988), pp. 37, 38.

360   The four loves of the soul do not exhaust the realm of human love. For example, the love of money or pizza is called concupiscence, defined by the craving for the pleasant that includes both the soul and the body. See Aquinas, *Summa Theologica*, The First Part of the Second Part, Question 30.

361   Mitsuko Uchida, Dame Mitsuko Uchida: Mozart Sonatas 545, 570, 576, 533/494.

362   Plato, *Symposium* in *The Collected Dialogues of Plato*, ed. Edith Hamilton and Huntington Cairns (Princeton: Princeton University Press, 1961), 192e.

363   Erich Fromm, *The Art of Loving* (New York: Harper, 2006 [1956]), p. 4.

364   Erich Fromm, *The Art of Loving* (New York: Harper, 2006 [1956]), p. 4., p. 51.

365   For a detailed discussion of the three kinds of friendship, see Aristotle, *Nichomachean Ethics*, Bks. VIII and IX.

366   Aristotle, *Nichomachean Ethics*, trans. Martin Ostwald (Indianapolis, IN: Bobbs-Merrill, 1962), Bk. VIII, line 1163b25.

367   See Aristotle, *Rhetoric*, Bk. II, Ch. 4, line 1381a.

368   In *The Bhagavad Gita*, the Sanskrit word that corresponds most closely to *agápē* is *tyāga*, the selfless action that has no desire for personal reward. See *The Bhagavad Gita*, trans. Eknath Easwaran with chapter introductions by Diana Morrison (Tomales, CA: Nilgiri Press, 1985), pp. 202-204 and 18.11.

369   Thomas Aquinas, *Summa Theologica*, II-II, Question 24, Article 2.

370   Sudhir Kakar, *The Inner World: A Psycho-analytic Study of Childhood and Society in India*, rev. ed. (New Delhi, India: Oxford University Press, 1981), pp. 80, 82.

371   John Bowlby, *A Secure Base: Clinical Applications of Attachment Theory* (London: Routledge, 1988), p. 61.

372   Erich Fromm, *The Art of Loving* (New York: Harper, 2006 [1956]), p. 4., p. 51., p. 39.

373   Billy Joel captures the adult desire for unconditional love in his "Just the Way You Are."

374   See "Dying in Your Mother's Arms." Found at https://www.nytimes.com/video/opinion/100000007249913/dying-in-your-mothers-arms.html?action=click>ype=vhs&version=vhs-heading&module=vhs®ion=title-area&cview=true&t=27

375   René Spitz, *The First Year of Life: A Psychoanalytic Study of Normal and Deviant Development of Object Relations* (New York: International Universities Press, 1965). See also Robertson, J., and J. Bowlby, "Responses of Young Children to Separation from Their Mothers," Paris: *Courr. Cent. Int. Enf*, 1952.

376   René Spitz, *The First Year of Life: A Psychoanalytic Study of Normal and Deviant Development of Object Relations* (New York: International Universities Press, 1965), p. 270.

377   See George E. Vaillant, *Triumphs of Experience: The Men of the Harvard Grant Study* (Cambridge, MA: Harvard University Press, 2015), pp. 44-51 and George E. Vaillant, "From emotionally crippled to loving personality," (Ted Talk, November 2014).

378   For an intercultural perspective on the inner monologue, see Sogyal Rinpoche, *The Tibetan Book of Living and Dying* (San Francisco: Harper, 1992), Ch. 5.

379   Barry Green, *The Inner Game of Music* (New York: Doubleday, 1986), p. 14.

380   Quoted by Mihaly Csikszentmihalyi, *Beyond Boredom and Anxiety* (San Francisco: Jossey-Bass, 1975), p. 39.

381   Quoted by Mihaly Csikszentmihalyi, *Beyond Boredom and Anxiety* (San Francisco: Jossey-Bass, 1975), p. 39.

382   Quoted by Mihaly Csikszentmihalyi, *Flow: The Psychology of Optimal Experience* (New York: Harper & Row, 1990), p. 53.

383   Doug Robinson, "The Climber as Visionary," *Ascent* **9**, (1969): 8.

384   Yvon Chouinard, quoted by Doug Robinson, "The Climber as Visionary," *Ascent* **9**, (1969): 6.

385   Bill Russell and Taylor Branch, *Second Wind: The Memoirs of an Opinionated Man* (New York: Random House, 1979), p. 157. Italics in original.

386   Chuang Tzu, *The Way of Chuang Tzu*, trans. and ed. Thomas Merton (New York: New Directions, 1965), p. 158.

387   Toshihiko Izutsu, *Toward a Philosophy of Zen Buddhism* (Boston: Shambhala, 2001), p. 207.

388   Eknath Easwaran, *Original Goodness: Eknath Easwaran on the Beatitudes of the Sermon on the Mount* (Tomales, CA: Nilgiri Press, 2014), p. 74.

389   Pseudo-Dionysius the Areopagite, *Mystical Theology*, trans. Editors of the Shrine of Wisdom (Surrey England: Shrine Of Wisdom, 1923), Ch. III.

390   Mihaly Csikszentmihalyi, *Flow: The Psychology of Optimal Experience* (New York: Harper & Row, 1990), p. 53.

391   Henri Poincaré, *Science and Method*, trans. Francis Maitland (Mineola, NY: Dover, 2003), p. 53.

392   Henri Poincaré, *Science and Method*, trans. Francis Maitland (Mineola, NY: Dover, 2003), pp. 53-54. Italics added.

393   Karl Friedrich Gauss, quoted by Jacques Hadamard, *The Psychology of Invention in the Mathematical Field* (Princeton: Princeton University Press, 1949), p. 15.

394   Fred Hoyle, "The Universe: Past and Present Reflections," *Annual Review of Astronomy and Astrophysics* **20** (1982): 24, 25.

395   Henri Poincaré, *Science and Method*, trans. Francis Maitland (Mineola, NY: Dover, 2003), p. 55.

396   G. H. Hardy, *Ramanujan: Twelve Lectures on Subjects Suggested by His Life and Work,* (Cambridge: Cambridge University Press, 1940), p. 9.

397   Maarten Schmidt, quoted by Mitchell Wilson, *Passion to Know* (Garden City, New York: Doubleday, 1972), p. 53.

398   Maarten Schmidt, quoted by Mitchell Wilson, *Passion to Know* (Garden City, New York: Doubleday, 1972),, p. 56

399   Mozart, quoted by Hadamard, *The Psychology of Invention in the Mathematical Field*, p. 16.

400   Czeslaw Milosz, "Catholic Education," in *Native Realm: A Search for Self-Definition* (Berkeley: University of California Press, 1981), p. 87.

401   For "The Dance" read by Tom Bedlam, see https://www.youtube.com/watch?v=wKaoXy2KaJU.

402   Theodore Roethke, *On the Poet and His Craft*, ed. Ralph J. Mills, Jr. (Seattle: University of Washington Press, 1966), pp. 223-24.

403   Plato, *Seventh Letter* in *The Collected Dialogues of Plato*, ed. Edith Hamilton and Huntington Cairns (Princeton, NJ: Princeton University Press, 1989), 344b.

404   Edward J. Larson and Larry Witham, "Leading scientists still reject God," *Nature* **394** (23 July 1998): 313.

405   "Why There Almost Certainly Is No God" is the central argument of Richard Dawkins, *The God Delusion* (Boston; Houghton Mifflin, 2006), pp. 111-159.

406   Fred Hoyle, *The Intelligent Universe: A New View of Creation and Evolution* (New York: Holt, Rinehart, and Winston, 1984), p. 19.

407   Richard Dawkins, *The Blind Watchmaker: Why the Evidence of Evolution Reveals a Universe without Design* (New York: Norton, 1986), p. 7.

408   Aquinas, *Summa Theologica*, First Part, Question 3.

409   St. Anselm, quoted by R. W. Southern, *St. Anslem: A Portrait in a Landscape* (Cambridge: Cambridge University Press, 1990), p.128.

410   Bertrand Russell, "My Mental Development," in P. A. Schilpp, *The Philosophy of Bertrand Russell* (New York: Harper & Row, 1963), vol. 1, p. 10.

411   Bertrand Russell, *History of Western Philosophy (New York: Touchstone, 1967), p. 586.*

412   Steven Weinberg, *The First Three Minutes: A Modern View Of The Origin Of The Universe* (New York: Basic Books, 1977), p. 154.

413   Augustine, *The Literal Meaning of Genesis*, trans. John Hammond Taylor (New York: Newman press, 1982), I, p. 185.

414   Fred Hoyle, "The Universe: Past and Present Reflections," *Annual Review of Astronomy and Astrophysics* **20** (1982): 16. Available http://calteches.library.caltech.edu/527/2/Hoyle.pdf.

415   See Steven Weinberg, "The Cosmological Constant Problems" http://ned.ipac.caltech.edu/level5/Weinberg/frames.html.

416   M. P. Hobson, G. P. Efstathiou, and A. N. Lasenby, *General Relativity: An Introduction for Physicists* (Cambridge: Cambridge University Press, 2007), p. 187.

417   John Archibald Wheeler, "Genesis and Observership," in *Foundational Problems in the Special Sciences,* ed. Robert E. Butts and Jaakko Hintikka (Dordrecht, Holland: Reidel, 1977), p. 18.

418   Blaise Pascal, *Pensées*, trans. A.J. Krailsheimer (New York: Penguin, 1995), [201].

419   See Leonard Susskind, Interview, Amanda Gefter, "Is string theory in trouble?", *New Scientist* (17 December 2005). Also see Steven Weinberg, "Living in the Multiverse," http://arxiv.org/abs/hep-th/0511037.

420   Steven Weinberg, "Physics: What We Do and Don't Know," *The New York Review of Books* (November 7, 2013).

421   Lawrence M. Krauss, *A Universe from Nothing: Why There Is Something Rather than Nothing* (New York: Free Press, 2012), p. 176.

422   Paul Steinhardt, "Big bang blunder bursts multiverse bubbles," *Nature* **510** (5 June 2014): 9.

423   Richard Feynman, Interview, in P. C. Davies and Julian Brown, eds. *Superstrings: A Theory of Everything?* (Cambridge: Cambridge University Press, 1988), p. 194.

424   Sheldon Glashow, with Ben Bova, *Interactions: A Journey through the Mind of a Particle Physicist* (New York, NY: Warner Books, 1988), p. 25.

425   Richard C. Lewontin, "Billions and Billions of Demons," *The New York Review of Books* (January 1997): 37. Italics added.

426   See Gell-Mann, "Let's Call It Plectics," *Complexity* **1** (1995/1996), no. 5.

427   Francis Crick, *The Astonishing Hypothesis*, (New York: Scribner's, 1994), p. 3.

428   See Richard Dawkins, *The Selfish Gene* (New York: Oxford University Press, 1976), p. 21.

429   A shorter version of this discussion of beauty in science appeared in Robert M. Augros and George N. Stanciu, *The New Story of Science: Mind and the Universe.*

430   Albert Einstein, "Autobiographical Notes," in *Albert Einstein: Philosopher-Scientist*, ed. Paul Schilpp (New York: Harper & Row, 1959), p. 33.

431   Stephen Daedalus, the protagonist of James Joyce's *A Portrait of the Artist as a Young Man*, argues that the three universal qualities of beauty in the arts are wholeness, harmony, and radiance, his translation of Aquinas' integritas, consonantia, and claritas.

432   Roger Penrose, "Black Holes," in *The State of the Universe*, ed. Geoffrey Bath (Oxford: Clarendon Press, 1980), p. 128.

433   Werner Heisenberg, *Physics and Beyond: Encounters and Conversations*, trans. Arnold J. Pomerans (Harper & Row, New York, 1971), pp. 68-69.

434   Henri Poincaré, *Science et Méthode* (Paris: Flammarion, 1949), p. 16. Our translation.

435   Werner Heisenberg, "The Meaning of Beauty in the Exact Sciences," in *Across the Frontier* (New York: Harper & Row, 1974), p. 167.

436   Albert Einstein and Leopold Infeld, *The Evolution of Physics* (New York: Simon & Schuster, 1938), p. 313.

437   Werner Heisenberg, "The Meaning of Beauty in the Exact Sciences," in *Across the Frontier* (New York: Harper & Row, 1974), p. 167.

438   John A. Wheeler, "The Universe as a Home for Man," *American Scientist* **62** (Nov.-Dec. 1974): 688.

439   Werner Heisenberg, *Physics and Philosophy* (New York: Harper & Row, 1958), p. 133.

440   George Thomson, *The Inspiration of Science* (Oxford: Oxford University Press, 1961), p. 18.

441   S. Chandrasekhar, "Beauty and the Quest for Beauty in Science," *Physics Today* **32** (July 1979): 29.

442   Francis Bacon, "Of Beauty" in T*he Essayes or Counsels, Civill and Morall*, ed. Michael Kiernan (Cambridge, Mass.: Harvard University Press, 1985), p. 132. Available http://www.bartleby.com/3/1/.

443   Leonard Bernstein, *The Infinite Variety of Music* (New York: Simon & Schuster, 1966), p. 198. Italics in the original.

444   The ten most beautiful experiments in physics, according to a poll of Physics World readers, are: (1) Young's double-slit experiment applied to the interference of single electrons (1961); (2) Galileo's experiment on falling bodies (1589); (3) Millikan's oil-drop experiment (1909); (4) Newton's decomposition of sunlight with a prism (1665-1666); (5) Young's light-interference experiment (1801); (6) Cavendish's torsion-bar experiment (1798); (7) Eratosthenes' measurement of the Earth's circumference (3rd century BC); (8) Galileo's experiments with rolling balls down inclined planes (1604, published 1638); (9) Rutherford's discovery of the nucleus (1911); (10) Foucault's pendulum (1851). Also, see Robert P. Crease, *The Prism and the Pendulum: The Ten Most Beautiful Experiments in Science* (New York: Random House, 2004).

445   Bertrand Russell, *The Autobiography of Bertrand Russell: 1872–1914* (Boston: Atlantic Monthly Press, 1967), pp. 37-38.

446   Sigmund Freud, *Civilization and Its Discontents*, trans. and ed. James Strachey (New York: Norton, 1961), p. 33.

447   Charles Darwin, *On the Origin of Species*, 6th ed. (London: Murray, 1872), p. 161. Available http://darwin-online.org.uk/content/frameset?viewtype=text&itemID=F401&pageseq=1.

448   Charles Darwin, *On the Origin of Species*, 6th ed. (London: Murray, 1872), p. 162.

449   Charles Darwin, *On the Origin of Species*, 6th ed. (London: Murray, 1872), p. 160.

450   Charles Darwin, *The Voyage of the Beagle* (New York: Dutton, 1967), p. 8.

451   Darwin, *On the Origin of Species*, 6th ed. (London: Murray, 1872), p. 429.

452   Fred Hoyle, *Highlights in Astronomy*, (San Francisco: Freeman, 1975), pp. 35-36.

453   Louis de Broglie, *Savants et Découvertes* (Paris: Albin Michel, 1951), pp. 378-379. Our translation.

454   James Watson, *The Double Helix* (New York: Mentor, 1968), pp. 131, 134.

455   Richard Feynman, *The Character of Physical Law* (Cambridge: M.I.T. Press, 1965), p. 171.

456   All the numbers for the properties of an electron in this paragraph are from Steven Weinberg, *Dreams of a Final Theory* (New York: Pantheon, 1992), pp. 114-115.

457   Steven Weinberg, *Dreams of a Final Theory* (New York: Pantheon, 1992), p. 115.

458   Albert Einstein, quoted by Banesh Hoffman and Helen Dukas, *Albert Einstein, Creator and Rebel* (New York: Viking Press, 1972), p. 18.

459   Edward Witten, Glenn Starkman, and Lawrence Krauss quoted by Dennis Overbye, "The End of Everything," *The New York Times*, Science Section, January 1, 2002.

460   Eugene P. Wigner, "The Unreasonable Effectiveness of Mathematics in the Natural Sciences," *Communications in Pure and Applied Mathematics* **13** (February 1960) and also in Wigner's collection of essays *Symmetries and Reflections: Scientific Essays* (Bloomington, IN: Indiana University Press, 1967).

461   P. A. M. Dirac, "The Evolution of the Physicist's Picture of Nature," *Scientific American* **208** (May 1963): 53.

462   Plato, Symposium in *The Collected Dialogues of Plato*, ed. Edith Hamilton and Huntington Cairns (Princeton, NJ: Princeton University Press, 1989), 211c. Italics in the original.

463   The conclusion God is intelligible, good, and beautiful, if read literally, is false. No formula, no set of words, no elaborate theology can contain God, the Unnamable. See Pseudo-Dionysius, *The Divine Names,* in *Pseudo-Dionysius: The Complete Works,* trans. Colm Luibheid (New York: Paulist Press, 1987), 596A and 872; and Pseudo-Dionysius, *The Mystical Theology,* in *Pseudo-Dionysius: The Complete Works*, trans. Colm Luibheid (New York: Paulist Press, 1987), 1033B and also Chapter 21.

464   Pseudo-Dionysius the Areopagite is the anonymous theologian of the late 5th to early 6th century whose works were erroneously ascribed to Dionysius the Areopagite, the Athenian convert of St. Paul mentioned in Acts 17:34.

465   Pseudo-Dionysius, *The Divine Names* in *Pseudo-Dionysius: The Complete Works*, trans. Colm Luibheid (New York: Paulist Press, 1987), 596A.

466   Pseudo-Dionysius, *The Divine Names* in *Pseudo-Dionysius: The Complete Works*, trans. Colm Luibheid (New York: Paulist Press, 1987), 872A.

467   Pseudo-Dionysius, *The Mystical Theology* in *Pseudo-Dionysius: The Complete Works*, trans. Colm Luibheid (New York: Paulist Press, 1987), 1033B

468   John of Damascus, quoted by Timothy Ware *The Orthodox Church* (London: Penguin, 1993), p. 63.

469   Peter of Damascus, *Twenty-Four Discourses* in *The Philokalia*, vol. III, ed. and trans. G. E. H. Palmer, Philip Sherrard, and Kallistos Ware (London: Faber and Faber, 1984), p. 255.

470   Maximos the Confessor, *Four Hundred Texts on Love* in *The Philokalia*, vol. II, ed. and trans. G. E. H. Palmer, Philip Sherrard, and Kallistos Ware (London: Faber and Faber, 1981), p. 64.

471   Maximos the Confessor, *Four Hundred Texts on Love* in *The Philokalia*, vol. II, ed. and trans. G. E. H. Palmer, Philip Sherrard, and Kallistos Ware (London: Faber and Faber, 1981), p. 69.

472   Gregory Palamas, *Topics of Natural and Theological Science and on the Moral and Ascetic Life: One Hundred and Fifty Tests* in *The*

Philokalia, vol. IV, ed. and trans. G. E. H. Palmer, Philip Sherrard, and Kallistos Ware (London: Faber and Faber, 1984), p. 382.

473   Thomas Merton, *New Seeds of Contemplation* (New York: New Directions, 1962), p. 132. Italics in original.

474   Aquinas, *Quaestiones Disputatae De Potentia Dei*, Q. VII: Article V.

475   Aquinas, *Quaestiones Disputatae De Potentia Dei*, Q. VII: Article V.

476   See Aquinas, *Quaestiones Disputatae De Potentia Dei*, Q. VII: Article V.

477   Pseudo-Dionysius, *The Mystical Theology*, 1048B.

478   Pseudo-Dionysius, *The Mystical Theology*, 1048A.

479   Pseudo-Dionysius, *The Mystical Theology*, 1033C, 1001A.

480   René Descartes, *Meditations on First Philosophy* in *The Philosophical Writings of Descartes Volume II*, trans. John Cottingham, Robert Stoothoff, and Dugald Murdoch (Cambridge: Cambridge University Press, 1984), p. 35.

481   Christopher Hitchens, "Introduction," *The Portable Atheist: Essential Readings for the Nonbeliever*, ed. Christopher Hitchens (New York: Da Capo Press, 2007), p. xxii.

482   See Plato, *Timaeus*, in *The Collected Dialogues of Plato*, ed. Edith Hamilton and Huntington Cairns (Princeton, NJ: Princeton University Press, 1989).

483   Anthony Meredith, "On the Life of Moses" in *Gregory of Nyssa: The Early Church Fathers* (New York: Routledge, 1999), 2.165.

484   Lucretius, *De Rerum Nature*, trans. Rolfe Humphries (Bloomington: Indiana University Press, 1969), p. 34.

485   Nicolaus Copernicus, *On the Revolutions of the Heavenly Spheres*, trans. R. Catesby Taliaferro in *Great Books of the Western World*, (Chicago: Encyclopedia Britannica, 1939), vol. 16, p. 508.

486   Nicolaus Copernicus, *On the Revolutions of the Heavenly Spheres*, trans. R. Catesby Taliaferro in *Great Books of the Western World*, (Chicago: Encyclopedia Britannica, 1939), vol. 16, p. 506.

487   Nicolaus Copernicus, *On the Revolutions of the Heavenly Spheres*, trans. R. Catesby Taliaferro in *Great Books of the Western World*, (Chicago: Encyclopedia Britannica, 1939), vol. 16, p. 103.

488   Francis Bacon, *The Advancement of Learning*, ed. William Aldis Wright (Oxford: Oxford University Press, 1963 [1605]), p. 45.

489   Francis Bacon, *The New Organon and Related Writings* (Indianapolis, IN: Bobbs-Merrill, 1960 [1620]), p. 23.

490   Patriotism—the love of place, countrymen, and local traditions—lasted for millennia, until replaced by nationalism, which we believe is a natural outgrowth of tribal life, instead of an invention of Western Europe. For a fuller discussion, see George Stanciu, "The Virtues of Patriotism, the Vices of Nationalism." May be found at https://theimaginativeconservative.org/2016/11/virtues-patriotism-vices-nationalism-george-stanciu.html

491   Thomas Jefferson, "Second Inaugural Address," March 4, 1805. May be found at https://avalon.law.yale.edu/19th_century/jefinau2.asp

492   John Locke, *Second Treatise of Government*, ed. C. B. Macpherson (New York: Hafner, 1980 [1690]), p. 66.

493   See Aristotle, *Nicomachean Ethics*, Bk. VIII, Ch. 1.

494   Galatians 5:14, 13. RSV.

495   Aeschylus, *Prometheus Bound*, trans. David Grene in *Aeschylus I: The Persians, The Seven Against Thebes, The Suppliant Maidens, Prometheus Bound* (Chicago: University of Chicago Press, 2013), lines 230-255.

496   Su Tung-po, *Remembrance*. May be found at https://mypoeticside.com/show-classic-poem-28763

497   Eccl 1:2-4. RSV

498   Quoted by Sushila Blackman, ed., *Graceful Exits: How Great Beings Die (Death stories of Hindu, Tibetan Buddhist, and Zen masters)*, (Boston: Shambhala, 2005), p. 7.

499   See Blaise Pascal, *Pensées*, trans. W. F. Trotter (Grand Rapids, MI: Christian Classics Ethereal Library, 2005), p. 30.

500   "The Buddha's First Sermon, Known as the Foundation of the Kingdom of Righteousness or the Setting in Motion of the Wheel of Dharma," in *Buddhism: A Religion of Infinite Compassion*, ed. Clarence H. Hamilton (Indianapolis, IA: Bobbs-Merrill, 1952), pp. 28-29 and "The Sermon at Benares," in *The Teachings of the Compassionate Buddha*, ed. E. A. Burtt (New York: New American Library, 1955), p. 30.

501   Robert Wright, *Why Buddhism is True: The Science and Philosophy of Meditation and Enlightenment* (New York: Simon & Schuster, 2017), pp. 6, 55.

502   Robert Wright, *Why Buddhism is True: The Science and Philosophy of Meditation and Enlightenment* (New York: Simon & Schuster, 2017), p. 55.

503   John von Neumann in conversation with Stanislaw Ulam, "Tribute to John von Neumann, 1903-1957," *Bulletin of the American Mathematical Society* **64** (May 1958).

504   Vernor Vinge, *The Coming Technological Singularity: How to Survive in the Post-Human Era.* May be found at http://edoras.sdsu.edu/~vinge/misc/singularity.html

505   Ray Kurzweil, *The Singularity Is Near: When Humans Transcend Biology* (New York: Penguin Books, 2006), p. 9.

506   Jaron Lanier, *You Are Not a Gadget: A Manifesto* (New York: Vintage, 2011), p. 29.

507   Yuval Noah Harari, *Homo Deus: A Brief History of Tomorrow* (New York: Harper, 2017), p. 28.

508   Sergey Brin, quoted by Tad Friend, "Silicon Valley's Quest to Live Forever," *The New Yorker* (April 3, 2017).

509   David Rieff, *Swimming in a Sea of Death: A Son's Memoir* (New York: Simon & Schuster, 2008), p. 38.

510   David Rieff, *Swimming in a Sea of Death: A Son's Memoir* (New York: Simon & Schuster, 2008), p. 39.

511   David Rieff, *Swimming in a Sea of Death: A Son's Memoir* (New York: Simon & Schuster, 2008), pp. 15, 13.

512   David Rieff, *Swimming in a Sea of Death: A Son's Memoir* (New York: Simon & Schuster, 2008), p. 61.

513   David Rieff, *Swimming in a Sea of Death: A Son's Memoir* (New York: Simon & Schuster, 2008), p. 17.

514   David Rieff, *Swimming in a Sea of Death: A Son's Memoir* (New York: Simon & Schuster, 2008), p. 150.

515   David Rieff, *Swimming in a Sea of Death: A Son's Memoir* (New York: Simon & Schuster, 2008), 173.

516   David Rieff, *Swimming in a Sea of Death: A Son's Memoir* (New York: Simon & Schuster, 2008), p. 171.

517   David Rieff, *Swimming in a Sea of Death: A Son's Memoir* (New York: Simon & Schuster, 2008), p. 163.

518   Plato, *Phaedo*, trans. Hugh Tredennick, in *The Collected Dialogues of Plato*, ed. Edith Hamilton and Huntington Cairns (Princeton: Princeton University Press, 1961), 67e.

519   Plato, *Seventh Letter*, trans. L. A. Post, in *The Collected Dialogues of Plato*, ed. Edith Hamilton and Huntington Cairns (Princeton: Princeton University Press, 1961), 343c.

520   Plato, *Phaedo*, 67c.

521   Plato, *Phaedo*, 84b.

522   Plato, *Phaedo*, 77d.

523   Plato, *Phaedo*, 117c-118.

524   C. S. Lewis, *The Problem of Pain* (New York: The Macmillan Company, 1962), Ch. 9.

525   Margaret Guenther, "God's plan surpasses our best imaginings," *Episcopal Life* (July/August 1993).

526   Oliver Sacks, *The Man Who Mistook His Wife for a Hat* (New York: Summit, 1987), p. 39.

527   Jerome Kagan, *Unstable Ideas: Temperament, Cognition, and Self* (Cambridge: Harvard University Press, 1989), p. 233.

528   Qi Wang and Jens Brockmeier, "Autobiographical Remembering as Cultural Practice: Understanding the Interplay between Memory, Self and Culture," *Culture & Psychology* (2002) **8**:52.

529   For a scientific study of implanted memories, see I. E. Hyman Jr. and J. Pentland, "The Role of Mental Imagery in the Creation of False Childhood Memories," *Journal of Memory and Language* (1996) **35** (2): 101–17.

530   Daniel Offer, Marjorie Kaiz, Kenneth L. Howard, and Emily S. Bennett, "The Altering of Reported Experiences," *Journal of the American Academy of Child and Adolescent Psychiatry* (June 2000), **39** (6): 735-42.

531   See Jonathan W. Schooler and Tonya Engstler-Schooler, "Verbal Overshadowing of Visual Memories: Some Things Are Better Left Unsaid," *Cognitive Psychology* **22** (1990): 36-71.

532   Hu Shih, quoted by Ambrose Yeo-chi King, "Kuan-hsi and Network Building: A Sociological Interpretation," *Daedalus*, **120** (Spring 1991): 65.

533   Richard E. Nisbett, *The Geography of Thought: How Asians and Westerners Think Differently . . . and Why* (New York, NY: Free Press, 2003), p. 51.

534   *Anatta-lakkhana Sutta: The Discourse on the Not-self,* in *In the Buddha's Words: An anthology of Discourses from the Pāli Canon,* trans. Bhikkhu Bodhi (Boston: Wisdom Publications, 2005), p. 342.

535   Walpola Rahula, *What the Buddha Taught* (New York: Grove Press, 1974), p. 37.

536   Burtt, ed., *Teachings of the Compassionate Buddha*, p. 113.

537   Pseudo-Dionysius, *The Divine Names* in *Pseudo-Dionysius: The Complete Works*, trans. Colm Luibheid (New York: Paulist Press, 1987), 596A.

538   Pseudo-Dionysius, *The Divine Names* in *Pseudo-Dionysius: The Complete Works*, trans. Colm Luibheid (New York: Paulist Press, 1987), 872A.

539   Gregory Palamas, *Topics of Natural and Theological Science and on the Moral and Ascetic Life: One Hundred and Fifty Texts*, in *The Philokalia*, Vol. IV, ed. and trans. G. E. H. Palmer, Philip Sherrard, and Kallistos Ware (London: Faber and Faber, 1984), p. 382.

540   Gregory Palamas, *Topics of Natural and Theological Science and on the Moral and Ascetic Life: One Hundred and Fifty Texts*, in *The Philokalia*, Vol. IV, ed. and trans. G. E. H. Palmer, Philip Sherrard, and Kallistos Ware (London: Faber and Faber, 1984), p. 383.

541   See Gregory Palamas, *Topics of Natural and Theological Science and on the Moral and Ascetic Life: One Hundred and Fifty Texts*, in *The Philokalia*, Vol. IV, ed. and trans. G. E. H. Palmer, Philip Sherrard, and Kallistos Ware (London: Faber and Faber, 1984), p. 382.

542   See Augustine, *City of God*, Bk. 14, Ch. 13.

543   Joseph Ratzinger, *Eschatology: Death and Eternal Life*, 2nd ed., trans. Michael Waldstein (Washington, D.C.: Catholic University of America Press, 1988), pp. 233-234.

George Stanciu

I am a Romanian gypsy from a long line of chicken stealers, fortunetellers, tax evaders, and draft dodgers. That is not such a bad heritage.

I received a Ph.D. in theoretical physics from the University of Michigan, and my experience teaching the Great Books at St. John's College (Santa Fe, New Mexico) is equivalent to a Ph.D. in the humanities. This almost unique background came about because from early on in my intellectual life, I pursued relentlessly fundamental questions.

While on a postdoctoral research fellowship at Los Alamos National Laboratory, I became bewildered by the hideousness of nuclear weapons and the nice guys that built them. I could not reconcile Hiroshima and the beauty of physics, nor could anyone else at Los Alamos. So, I decided to take what I thought would be a brief excursion in the humanities. I had the good fortune to study the humanities by co-teaching seminars on the Great Books at St. John's College in Santa Fe, New Mexico.

I left St. John's College to accept the position of Academic Dean of Magdalen College (Warner, New Hampshire). There, I founded a program that employs primary texts and the Socratic method of instruction to explore the first principles of nature and the fundamental questions of human life.

Thus, in one way or another, my entire intellectual life has been spent trying to make sense of human existence in dark times.

I am the Academic Dean Emeritus at Magdalen College of Liberal Arts in Warner, New Hampshire, and the co-author of *The New Biology* and *The New Story of Science*. I am also the author of *The Great Transformation: How Contemporary Science Harmonizes with the Spiritual Life*.